BUSINESS PROCESS MANAGEMENT APPLIED

Creating the Value Managed Enterprise

Charles Poirier
Ian Walker

APICS.
THE EDUCATIONAL SOCIETY
FOR RESOURCE MANAGEMENT

Copyright ©2005 by Charles C. Poirier and Ian Walker

ISBN 1-932159-33-9

Printed and bound in the U.S.A. Printed on acid-free paper
10 9 8 7 6 5 4 3 2 1

Library of Congress Cataloging-in-Publication Data

Poirier, Charles C., 1936–
  Business process management applied : creating the value managed
enterprise / by Charles Poirier and Ian Walker.
    p. cm.
  Includes bibliographical references and index.
  ISBN 1-932159-33-9 (hardback : alk. paper)
  1. Business logistics—Management. 2. Process control. 3. Value added.
4. Corporate profits. 5. Industrial management. I. Walker, Ian, 1953– II.
Title.
  HD38.5.P6373 2005
  658.5—dc22
                                                    2005011230

   Direct all inquiries to J. Ross Publishing, Inc., 6501 Park of Commerce Blvd., Suite 200,
Boca Raton, Florida 33487.

Phone: (561) 869-3900
Fax: (561) 892-0700
Web: www.jrosspub.com

DEDICATION

For Carol
   —Chuck

For Faith
   —Ian

TABLE OF CONTENTS

Preface vii

Acknowledgments xiii

About the Authors xv

About APICS xvii

Web Added Value™ xix

Chapter 1.   The Road to the Value-Managed Enterprise 1

Chapter 2.   Business Process Management as the Enabling Ingredient ... 33

Chapter 3.   The Process-Based Profit and Loss Statement
             and Balance Sheet 55

Chapter 4.   Getting Beyond the Supply Chain Roadblocks 79

Chapter 5.   Moving Effectively to Advanced Levels of Progress 101

Chapter 6.   Achieving Dominance at the Higher Levels of Progress 119

Chapter 7.   Aspirations and Realities of Level 5 155

Chapter 8.   Creating the Intelligent Value Network: A Blueprint for
             Level 5 173

Chapter 9.   Using Process Simulation to Minimize the Risk 189

Chapter 10. Steps Forward 209

Bibliography 227

Index 229

PREFACE

The search for savings and improvements within a business is a never-ending quest. It occupies the minds and energies of virtually every business, at all time, in every type of enterprise. Throughout the world, companies have launched a myriad of efforts to meet this need and to transfer the results of continuous improvement to the profitability of their businesses. During the past two decades, in particular, an ever-increasing number of firms have opted to bring these efforts under an umbrella improvement technique dubbed supply chain. With that orientation, the effort becomes focused on the end-to-end process steps that occur from wherever is the appropriate place to start the chain of supplies or services — through manufacturing, production, or delivery — to final consumption by a business customer or end consumer. When returns are considered as well, this focus is truly across the total processing that could occur from beginning to end of an extended enterprise business system.

Unfortunately, the actual results from supply chain efforts have been widely varied. Some firms, across all parts of the globe and in many sectors, have been successful in transitioning their continuous improvement efforts into an appropriate form of supply chain focus and can document significant savings — from three to eight points of new profit margin. Similar results seem to be elusive for other companies, even after a score of years of trying. Surveys by Computer Sciences Corporation (CSC), *Supply Chain Management Review* magazine, and other qualified and respected organizations verify the possibility of using supply chain as an effective tool for further enhancement of virtually any business improvement effort. At the same time, these studies show that the majority of firms have not reaped the full benefits found by the leaders in an industry, as the followers remain bogged down in the preliminary levels of their supply chain effort or stuck with cultural imperatives that limit enterprise collaboration.

Much work apparently remains, to get the intended results reflected in operating income, in particular using the benefits of supply chain not only to reduce costs but to increase revenues.

THERE ARE MISSING INGREDIENTS

The ability to attain the desired results should not be that difficult, based on the documentation reported, much of which we will reference and consider. Two ingredients seem to be missing. First, there is an absence of a concerted effort across the full business, focused on the type of improvements that will enhance customer satisfaction while bringing verifiable savings to the firm extending the effort — an effort that brings attention to the importance of the top line (revenues) as well as the bottom line (earnings). Second, there is an equal absence of a means to access the various databases involved in inter-enterprise processing, for the purpose of exchanging knowledge vital to any effort aimed at optimization across the end-to-end value chain. Most firms seem to be unaware of the advantages offered by business process management (BPM) and its enabling language and system (BPML and BPMS) to facilitate such data sharing.

As supply chain efforts, designed to identify and secure the full benefits for a firm, have matured, a framework for moving to the most advantageous position has emerged and guided many businesses to very impressive improvements. Leading companies appear to have advanced beyond industry competitors, as they are better able to implement against this framework, which requires a substantial commitment to using collaboration and technology through external alliances. They do so with a greater understanding of the importance of bringing a deeper focus to the key process steps in what becomes an *intelligent value network,* connecting multiple enterprises with their customers. Such a network is oriented toward the acquisition, management, and integration of customer information to create a differentiating customer value proposition, a commodity missing in action in most supply chain efforts.

In terms of profitability, positions achieved by industry-leading organizations, such as Colgate-Palmolive, Dell, Intel, Nestlé, Nike, Procter & Gamble, Tesco, Toyota, and Wal-Mart, have brought anywhere from one to three (for a three-year concerted effort) to five to eight (for a five- to ten-year effort) points of new profit to their bottom lines. The laggards have failed to generate any appreciable return on investment (ROI) that can be documented. The leaders have discovered the advantages offered by moving their supply chains into a position of having superior capabilities, gained through greater access to knowledge across what becomes an intelligent value chain network. The laggards continue to hammer away at cost reduction, primarily through continued pressure on suppliers for price concessions and attempts to take business processes to offshore locations with much lower labor rates.

From another aspect, distancing an individual business from its competitors in areas of importance in a market has long been the goal of most enterprises. With the breakthrough possibilities introduced with BPM techniques, a new vista appears. The chances to optimize operating conditions and extend market leadership have become concurrent possibilities. Gaining a dominant market position through the application of collaboration and technology focused on customer satisfaction, the key ingredient of the intelligent value network, has become a viable option — for those businesses willing to overcome normal cultural barriers and the traditional unwillingness to work cooperatively with external resources. The final part of the effort, bringing clear documentation of the savings into view, takes the journey full circle and substantiates the position gained by the leaders, often one of market dominance.

OUR PURPOSES AND EXPERIENCES VALIDATE THE SAVINGS

In this book, we will explore the maturity of supply chain efforts and document the specific savings that are possible from a sustained effort. We will go further and explain how a firm can accomplish the most difficult part of the effort, which is to validly track the actual improvements and monetary savings to the profit and loss (P&L) statement. Along the way, actual case studies and examples of successful implementations will be documented to substantiate the arguments being presented. Throughout the text, generic and specific models will be used, from many industries and companies, to illustrate how firms can develop specific solutions that will distinguish their customer offerings. Several collaborating organizations, in particular ArvinMeritor™, Lanner, Yantra, and Stratascope, have supplied details of their experiences in bringing leadership positions to firms willing to follow the roadmap and to make certain the ROI meets difficult parameters set on any investment of the type needed to carry out an extended enterprise supply chain effort. This information will be included as well, to bring a flavor for actual potential to the various chapters.

We intend to help the reader understand how process improvement can add value for any firm of any size in any business, how BPM becomes the key enabler in the process, and the way to track those savings to verify the ROI achieved. In a step-by-step manner, we will outline how progress is made with the supply chain framework and substantial improvements are achieved. A simulation technique will be explained as well, to help any firm experiment with the inherent concepts without placing the business at great risk during the necessary piloting to test various alternative approaches and solutions.

THE FRAMEWORK IS BASIC TO NETWORK DIFFERENTIATION

The book will introduce the roadmap for achieving success by relating specific process improvements for specific savings and new values. It begins with a guiding framework and a presentation of the underlying architecture, including the basic elements: optimizing the extended enterprise, applying BPM tools and techniques, and bringing value to all constituents of the network enterprise, especially the customers and end consumers. The result is a means for any company to establish the path to creating and sustaining a truly viable, linked and optimized intelligent business network, which delivers greater value than any competing network.

Any reader concerned with process improvement and the means to help a business organization optimize its supply chain efforts and increase shareholder value will benefit from reading this book. Such an effort only succeeds when the network, in a way that differentiates the effort from competing groups, satisfies the final customers and consumers better than competing networks. The book combines our supply chain expertise and firsthand knowledge to explain the means of enhancing business improvement efforts, verified by action studies across a wide range of industries. The chapters will guide any business organization through the supply chain evolution, past the areas that typically obstruct progress, and further — to as close to optimum operating conditions as possible and to levels of satisfaction not normally attained.

This book is for any reader who wants to understand what can truly come from a concerted effort to reach the highest level of progress important to a firm and its customers, how to track the improvements to meaningful benefits for both parties, and to see the results on a documented P&L statement.

The book begins with an introduction to the extended enterprise concepts and a five-level maturity model for tracking the route to advanced business performance. A description is provided of the basic building blocks that lead to the radical concepts behind our arguments. A generic business model is also introduced, which parallels the traditional SCOR® model developed by the Supply-Chain Council, but covers new products and services, how to execute best supply chain processes, and establishes the supporting enterprise processes in human resources, finance, and information technology. New technologies that form the backbone of BPM systems are considered as they open the door for business users to take back ownership of their processing from the information technology department. With an understanding that people are as critical to business success as technology is, the ability to continuously reengineer critical processes in real time gives leaders a significant advantage.

An important distinction develops as the authors introduce the idea that results from successful supply chain and networked processing efforts should be tracked directly to the financial performance of the firm and its business allies. Once the link is made between crucial business processes and the financial management of the enterprise, the real opportunities for increasing value can be identified. This concept is encapsulated in the transition values for each major process in the maturity model, enabling the reader to create a first-cut value-based transition plan.

Getting beyond the supply chain roadblocks comes next, as we detail the changes necessary for successful movement to the advanced levels of progress and how a firm can transition over the normal cultural barriers. Process improvement is finely mapped, using advanced modeling tools, which can be instantly deployed in the business. Citing results from surveys conducted by CSC and *Supply Chain Management Review* magazine, the authors explain how and why most firms remain bogged down in levels 2 and 3 of the maturity model and what is required for further advancement. The story continues with details on how some firms have successfully advanced to level 3 and beyond, with the help of willing and trusted business allies. The pitfalls are overcome and greater value added, particularly in the eyes of the key customers, as the supply chain becomes an intelligent value network, composed of linked businesses managing the processes across an extended enterprise. In particular, attention is given to the extra values that can be achieved in such areas as reaction time reduction and lesser order-to-cash cycle times.

Reaching market dominance at the higher levels of progress becomes a reality as we study how a firm should decide what level it should strive to achieve and why. Here the authors look at the most important part of the progression, attaining a differentiating position in the eyes of the key customers, and describe the area in which a distinguished position can be achieved. Since conditions of this part of the progression are won at a price, we take care to explain how investment and execution decisions are made in conjunction with the desires and needs of value chain partners. The book then explores the dreams and promises contained in attaining the highest level of progress in the maturity model, the level 5 realm of full network connectivity. Since few companies comprehend the concepts and meaning behind such connectivity, we demonstrate, through the help of our collaborating partners, how co-managed, intelligent value networks are the hallmark characteristic of the few organizations that reach this level. Processes flow, achieved seamlessly from end to end of the network, and daily planning and scheduling decisions take account of information from members across the extended enterprise. BPM systems become the enabling technology link.

The penultimate chapter looks at one of the latest developments in process management: the ability to apply process simulation to test alternative scenarios and advanced supply chain designs. The coming of BPM systems is being seen as an enabler to off-line testing of process redesigns, before letting them loose across the business enterprise. This step is critical in a business that has the ability to deploy process change instantly. By conducting simulated dry runs of the redesigned process against a test environment, the risks are minimized and the need to make important changes is highlighted. The concluding chapter summarizes the main themes of the book and the authors' conclusions, as a framework for moving forward is described.

ACKNOWLEDGMENTS

As we attempt to recognize those who provided assistance with this book, there are more names than we could possibly list. The effort has been the culmination of years of testing, application, and a few retraced paths, along a pathway that has taken us around the globe many times. Crediting the sources, which have added value along the way, has been done as carefully as possible throughout the text. If a particular individual or organization has been overlooked, it is purely unintentional.

We would especially like to thank Michael Bauer at CSC, Bruce Brien at Stratascope, Geoff Hook at Lanner, and Bob Steele at Yantra, for their significant input to the story.

Appreciation also goes to our usual cast of supporters, including Alex Black, George Borza, Ron Brown, Simon Buesnel, Steve Caulkins, Chet Chetzron, Chris Deacon, David Durtsche, Lynette Ferrara, Drew Gant, Steve Goble, Francis Hayden, Bill Houser, Dave Howells, Marty Jacobsen, Gary Jones, Chris Lennon, Douglas Neal, Joel Polakoff, Mike Powell, Frank Quinn, Steve Reiter, Elsbeth Shepherd, Steve Simco, George Swartz, Chuck Troyer, and Chuck Wiza.

ABOUT THE AUTHORS

Chuck Poirier is a recognized authority on process improvement, supply chain management, e-business techniques, and the collaborative use of technology around the world. He has written nine business books, six of which are related to improving supply chain processing. His work has been translated into nine languages. In addition to his work as a partner in the CSC Supply Chain Practice, he is a frequent presenter at national and international conferences and industry meetings. With more than 40 years of business experience, including senior-level positions, and the extensive research conducted for the writing of his many books, white papers, and position documents, Mr. Poirier is comfortable before any audience seeking help with its value chain networks. He has assisted many firms in a variety of industries to establish the framework for their supply chains and to find the hidden values across the collaborative networking that can be established. His advanced techniques have become a hallmark of firms seeking the most benefits from cross-organizational collaboration.

Ian Walker has many years of experience in industry, both as a line manager and as a leader of major change programs. He is an associate director with CSC Consulting, based in the United Kingdom. A recognized expert in supply chain management, he has a broad understanding of the complex issues involved and a strong ability to articulate possible avenues of improvement. Mr. Walker specializes in supply chain strategy, the application of leading-edge thinking, and the use of information technology to support business processes within and across extended enterprises. Throughout an impressive consulting career, he has worked with a wide range of clients in many industries, from retailing through engineering to process industries such as chemicals and pharmaceuticals. Through the use of innovative analytical and business change techniques, applied to many aspects of supply chain, he has helped companies achieve significant improvements to operational and financial performance. He is a frequent speaker and contributor to journals and media communication vehicles.

ABOUT APICS

APICS — The Educational Society for Resource Management is a not-for-profit international educational organization recognized as the global leader and premier provider of resource management education and information. APICS is respected throughout the world for its education and professional certification programs. With more than 60,000 individual and corporate members in 20,000 companies worldwide, APICS is dedicated to providing education to improve an organization's bottom line. No matter what your title or need, by tapping into the APICS community you will find the education necessary for success.

APICS is recognized globally as:

- The source of knowledge and expertise for manufacturing and service industries across the entire supply chain
- The leading provider of high-quality, cutting-edge educational programs that advance organizational success in a changing, competitive marketplace
- A successful developer of two internationally recognized certification programs, Certified in Production and Inventory Management (CPIM) and Certified in Integrated Resource Management (CIRM)
- A source of solutions, support, and networking for manufacturing and service professionals

For more information about APICS programs, services, or membership, visit www.apics.org or contact APICS Customer Support at (800) 444-2742 or (703) 354-8851.

Web
Added
Value

Free value-added materials available from
the Download Resource Center at www.jrosspub.com

At J. Ross Publishing we are committed to providing today's professional with practical, hands-on tools that enhance the learning experience and give readers an opportunity to apply what they have learned. That is why we offer free ancillary materials available for download on this book and all participating Web Added Value™ publications. These online resources may include interactive versions of material that appears in the book or supplemental templates, worksheets, models, plans, case studies, proposals, spreadsheets and assessment tools, among other things. Whenever you see the WAV™ symbol in any of our publications, it means bonus materials accompany the book and are available from the Web Added Value Download Resource Center at www.jrosspub.com.

Downloads available for *Business Process Management Applied: Creating the Value Managed Enterprise* consist of slides and white papers that illustrate an intelligent value chain business architecture, business process map and simulation model, procurement and strategic sourcing model, and the benefits derived from progression to each new level in the supply chain maturity model.

THE ROAD TO THE VALUE-MANAGED ENTERPRISE

As supply chain efforts have matured around the world, a clear distinction has appeared. Businesses that embrace the inherent concepts are opening a serious gap over less able competitors. They do so by using advanced techniques — primarily adopting external collaboration and the use of technology enablers across linked organizations — to focus not only on cost improvement but revenue enhancement. The distinction that can be developed has been used by such leaders as Wal-Mart, Procter & Gamble, Toyota, Intel, Dell, Cisco Systems, Tesco, and Nike, to open up commanding leads in their industries. These organizations and others are moving rapidly toward what becomes a value-managed enterprise and a dominant position in an industry, while many of their competitors remain bogged down in the early levels of their supply chain progression.

The leaders have nurtured the advantages offered by moving their supply chains into a position of having superior capabilities, gained through greater access to knowledge across what becomes an intelligent value chain network. The difference for these businesses can be a doubling of earnings per share. The laggards and followers tend to keep their focus on internal improvement only, particularly the never-ending quest for lower costs of operations. While we accept that this cost orientation will hardly cease to exist, the contemporary view holds that there must be an equal and pervasive effort directed at distinguishing the firm in the eyes of the most important customers and end consumers so that new and profitable revenues are generated. Moreover, there must be a methodology in place to track the claimed improvements to the profit and loss statement, and there should be documented benefits for those members of the network that assist in the improvement process.

To facilitate our presentation and to establish a framework for understanding how a business can make progress toward the desired position with a supply chain effort and within an industry, we will use a familiar maturity model as a guide for the discussion. The purpose is to introduce the concept of supply chain optimization and how companies can approach their desired status and to apply a five-level progression model to explain the route to advanced business performance while calibrating the results. From this beginning position, we will present the process-enabled matrix, to explain how a firm can add the most benefits from improving the process steps in the intelligent value network. More importantly, we will describe how to trace the improvements to financial advantage — for the firm involved and its business allies. In later chapters, a simulation technique will also be discussed, so the interested firm can experiment with the techniques described without incurring undue risk. Throughout the text, actual case examples will be used to give the concepts a flavor of reality.

CONCEPTS BEHIND SUPPLY CHAIN MATURITY

In Figure 1.1, a supply chain maturity model is used to describe the typical progression, through which a firm evolves on its way to the most desired advanced level of implementation. Many firms in many countries have used this model to understand the logical progression of a supply chain effort. The model is also useful to calibrate a firm's position as it moves forward and to determine to what level of progress the company should aspire, especially when the model is extended to the various business functions. Implicit in the use of the model is the understanding that a firm must progress through the levels, none of which can be skipped, although some firms might have business units with footprints in various levels at the same time.

To begin, a firm typically launches its supply chain effort by bringing whatever existing improvement effort is being pursued under an umbrella type of orientation focused on the end-to-end processing that constitutes the organization's supply chain. In this first level of the progression, the firm begins to focus on functional processes, particularly sourcing and logistics. In addition to using these two areas of attention to gather early improvement and quick profit gains, a secondary goal is to bring enterprise integration into focus as an objective — within the organization. That means most companies find, as they begin integrating their existing improvement efforts with the overall attention to supply chain, that there is a substantial amount of internal resistance to any type of information sharing or best practice integration inside the four walls of

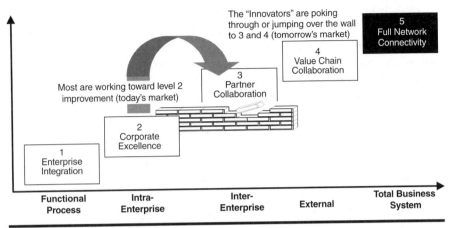

Figure 1.1.   The Supply Chain Evolution

the business. The so-called stovepipe or silo mentality is a serious reality for companies starting into the supply chain evolution.

As will be documented, there generally are significant savings generated during this first part of the evolution. Supply bases are reduced through techniques designed to segregate suppliers, based on categories of purchase, amount of purchase, quality, and factors of importance to the firm. With a lesser number of sources providing larger volumes, prices are reduced and added features introduced into the relationship. Typical benefits add one or two points of profit, because the amount of total purchases is often as much as 40 to 50% or more of the costs of goods or services sold.

Concurrently, most firms begin paying serious attention to how purchased goods are brought into the supply chain processing, how material and supplies are handled internally, and how finished products and goods are delivered to the next company in the linked processing. That means a focus is brought to inbound freight, internal material processing and manufacture, outbound freight, and the warehousing and delivery mechanisms necessary to support the logistics system. In the first level, the bulk of concern is brought to bear on internal processing and outbound freight. Inbound freight is generally held for later attention. There is a typical resistance to considering turning any part of the freight handling over to other parties, but as a firm completes this level, there is generally some form of outside help brought in to take responsibility for at least a portion of the transportation function. Most firms find one-half to a full percent of profit here, as they reduce the cost per mile for transportation, begin reducing warehouse space to what is actually needed, and transfer ownership of some or all of the assets involved to a third party adept at combining load requirements over the accumulated asset base for more effective utilization.

There is not much more to the first level, other than the fact that many firms begin looking for a model to guide the effort, with most selecting the SCOR® model, endorsed by the Supply-Chain Council. From our experiences, most businesses modify that model to suit their needs and introduce other features of importance to their particular industry. Virtually all have used the model described in Figure 1.1 to track their progress to higher levels of accomplishment. Detailed discussions of the various levels we are considering can be found in *Advanced Supply Chain Management* by Poirier (1999) and *E-Supply Chain* by Poirier and Bauer (2000).

In level 2, the business continues on its intra-enterprise track, but now the focus moves to attaining some degree of optimization across the internal processing occurring within the four walls of the business. Sourcing and logistics move to a higher level as the firm begins more intense work with the key suppliers and transportation providers. Planning, as accented by the SCOR® model, becomes an important effort. A supply chain infrastructure begins to appear during this level, usually with a professional supply chain manager being introduced to the organization.

Since planning now assumes importance, order management takes on relevance as a key factor. Unfortunately, virtually every study we have conducted in this area unearths a problem with the data. As orders, both with suppliers for purchased goods and services and with customers for finished products and services, are analyzed, we typically find that 30 to 40% contain one or more errors. This range may seem high, but when the analysis is extended to determining how the orders are placed or received, we find that most companies still rely heavily on telephone, facsimile, and mail for order entry and order processing. In spite of the amount of order management that has moved to electronic data interchange and e-commerce, the overwhelming amount of order processing is still done in a manual fashion, with many implicit errors. Establishing an order system that eliminates these errors and mistakes is crucial to introducing a planning system that has any degree of reliability and becomes a requisite for level 2 progress.

As planning is pursued, some form of sales and operations planning (S&OP) is generally introduced to require the firm to have a formal process for analyzing the orders, the production process steps, and the final delivery to customers. Forecast accuracy becomes an internal issue, as we find that most level 1 firms operate in an area of about 40% or less forecast accuracy. With the application of special algorithms or mathematical models, that figure might rise to about 50% in level 2. As S&OP matures, and key customers become involved in the level 3 processing, the accuracy of the demand information improves and firms do move as high as 60% accuracy. These general ranges are not absolute, but are based on our experiences with numerous firms across many industries.

In the second level, companies also turn their attention to the matter of inventory management. Much effort is expended to reduce the amount of cash tied up in inventories, so a one-time improvement can be made to working capital, and the annual carrying costs of excess inventories are reduced. Our experience shows, however, that some of the improvements made are real, while much of the effort results in a sort of shell-game maneuver. This means we find that a large amount of the inventory is simply moved upstream — toward obliging suppliers — rather than taken out of the supply chain. Results of many studies have indicated that firms did, indeed, remove a significant amount of inventory during their level 2 efforts, but much of it was absorbed by suppliers willing to hold the inventory until needed or to retain ownership until the point of actual use. Assuming the suppliers are smart enough to cover their carrying costs in the total price for the delivered supplies, it becomes a game of moving the assets and costs, rather than eliminating the need for extra inventory and safety stock.

Toward the conclusion of level 2, a supply chain organization will be firmly established, with a designated leader. Collaboration between this leader and the information technology group, and particularly the chief information officer, will be emerging. Internal resistance to cooperative effort will be subdued. The firm will be leveraging its full buying, manufacturing, storage, and delivery capacity across all business units, and there will be an intranet of information sharing at work, to supply the various functions and parts of the business with vital data on the processing taking place. Transportation management systems and warehouse management systems will be appearing as the practitioners begin to apply technology as an enabler to the desired process improvement. As indicated in the figure, most firms are still working on level 2 improvements in today's markets.

A cultural barrier exists between levels 1 and 2, as indicated by the brick wall in the figure. Most businesses are very comfortable continuing to work on internal excellence, particularly the never-ending quest for cost reduction. They resist outside help as an indictment that they do not have the internal expertise to solve each and every supply chain issue or problem encountered. Many also resist references to best practices that could be viewed and adopted outside of their four walls, and they especially oppose sharing any information that is considered proprietary with any external sources, even if it could lead to further improvement. Overcoming this barrier takes a forceful hand at the top of the organization and generally comes when one business unit leader takes a part of the firm into level 3 and beyond, proving the value of working collaboratively. A final requirement for completing this level is to form an alliance between the supply chain organization and the information technology group, so there is a joint effort to improve processing and to bring the necessary enabling technology into play. As we will document shortly, most firms have difficulty meeting the requirements for getting across this internal barrier.

In level 3, the firm begins to very selectively choose business allies to advance the supply chain improvement effort on a network basis. That means an inter-enterprise view is brought to the end-to-end processing, and partner collaboration is considered as an important business strategy. The idea is to have the most competent constituent of the extended enterprise assume responsibility for the process steps for which it is most capable of performing. Forecasting and planning go through another improvement cycle as companies begin to collaborate to match actual demand signals with production and delivery capacity. A form of advanced planning and scheduling is typically introduced in this level. The concepts are basic, but implementation is difficult, as many companies resist turning over responsibility for any processing to external organizations, even when the data verify the higher capability. They also want to hold back the sharing of vital information, until there is a better understanding of what can and should be shared for mutual advantage.

Most companies start their network formation by working closely with a few suppliers. They begin sharing a bit of previously sacrosanct data, with the intention of finding the hidden savings that have been eluding the firms, in spite of years of purchasing and supply relationships. Now the process maps describing the supply chain steps are viewed with an eye toward how the business allies can find areas of mutual benefit. Partnering diagnostics come into play as the allies seriously analyze what takes place within the process steps, particularly at the points of handoff. Frequently using reengineering techniques, the allies develop improved process steps, often applying enabling technology to improve the transfer of critical knowledge. Results generally provide a higher level of improvement for each of the participating parties. The key in this area is to make certain that the discovered savings are shared in some way between the parties to the effort.

Distributors can play a key role in this area, as they work more closely with the manufacturers to get the products and services to the end customer in the most effective manner, with the appropriate amount of inventory. The results of the survey to be cited verified that some of these distributors have made substantial progress with their supply chain efforts and have advanced solidly into level 3 and beyond. As the internal house begins to get in order, and key distributors are made a part of the network, most firms then turn to a few key customers and the analysis of the linked process steps continues, with the focus moving toward optimization across the total network, not just the internal processing.

In level 4, which includes a small percentage of the total companies pursuing supply chain optimization, we truly find the advanced supply chain management

(ASCM) efforts. Now the supply chain constituents are working closely together, sharing knowledge across a communication extranet and collaborating on how to increase revenues and better utilize combined assets, as well as finding further cost and service improvements. A new set of metrics is introduced to measure performance, with most of the measures focused on customer or consumer satisfaction. This is truly the realm of the intelligent value network we will describe and where the savings can reach as high as five to eight points of new profit.

The fifth and final level of the evolution is darkened in the figure, because it is more theoretical than actual, with only a handful of firms achieving full network connectivity. That term is used to describe a condition in which the linked organizations are sharing virtually all of the important data electronically and are working together through some form of cybercommunication system. This is the domain of high-technology companies (Intel, Cisco, Hewlett-Packard, IBM), a few leading-edge consumer goods providers (Colgate-Palmolive, Procter & Gamble, Nike, Dell, GE, Toyota), and a select group of experimenters trying to determine the actual value of direct knowledge sharing.

This framework will be used throughout the text, as we describe how a business organization can move carefully and effectively across the five levels and attain the highest position of importance to the firm, its market, and its customers. For now, the maturity model should be applied to consider how companies are making progress across the described evolution and to determine the values for moving to higher levels of performance. As we progress together, we will also show what the documented results by level have been.

CONFIRMED RESULTS VALIDATE THE OPPORTUNITY

During the summer of 2004, Computer Sciences Corporation (CSC) and *Supply Chain Management Review* conducted a survey among readers of the magazine and selected business organizations known to have some form of ASCM effort under way. The feedback came from 209 companies, of which 115 were corporations or independent businesses; 58 were divisions, wholly owned subsidiaries, or strategic business units; and 33 were group or multidivision organizations. The industries represented in the survey included aerospace and defense, automotive, chemicals, consumer goods, discrete manufacturing, food service, government, healthcare, high technology, mining, oil and gas, process manufacturing, professional services, publishing and printing, retail, telecommunications, utilities, and wholesale distribution; 23 firms listed their business as "other."

The survey was intended to find answers to such questions as:

- Where have supply chain efforts gone?
- What are the documented results?
- What are the leaders doing?
- What variation exists across industries?
- What is the link with technology?

In summary, the results provided the following answers:

- Most businesses, regardless of industry, reported that their results positioned them predominantly in levels 2 and 3, with evidence that most companies tend to bog down somewhere in this area; 91% of the respondents indicated they were no further along than level 3.
- Most of the documented progress, as expected, was reported in the areas of sourcing and logistics, followed by planning and inventory management. Those responding rated progress with other areas of importance much lower.
- Collaboration with external business allies remains an elusive concept for most companies, as there was clear evidence that this concept was not generally being applied.
- Reported cost savings ranged from 1 to 20% or more of *identified supply chain costs*. There appeared to be a disparity in what is included in supply chain costs, as the amount varied from a few percent to as much as 50% of total revenues, indicating that there still is some confusion concerning what should be included in supply chain costs.
- High technology, telecommunications, and wholesale delivery were the industries where those responding gave themselves the highest level of progress ratings.
- The link with technology was indelible, meaning higher levels of progress could not be attained without enabling technology, but there was sufficient information in the responses to indicate that there was far more technology applied to finding solutions and improvements than there were documented results as a result of those efforts. We interpret this finding to mean that there has been a tendency to try to apply technology solutions before the processing has been properly improved and then enhanced with technology.
- Revenue increases were also reported, ranging from 1 to 20% or more, but there was less certainty of the results here, as a significant percentage of those responding indicated they were not sure of revenue increases.

Our general conclusion was that supply chain efforts are alive and well, with supply chain now accepted as a major business improvement technique. Much more work remains to be accomplished, however, to get firms out of the gridlock occurring at levels 2 and 3 and moving, with the help of willing and trusted business allies, into the advanced levels of the progression. In the following chapters, we will describe some viable techniques and document the improvements made by some specific businesses to help in that quest.

There were several other important points that emerged from the survey. When we asked if the CEO considers supply chain management to be a source of competitive advantage, the replies were 75% positive, indicating the concept is now accepted at the most senior level of the business for most organizations. When we inquired as to what functions/costs were included in a supply chain organization, there were two categories to the responses. The major functions included, in ranked order:

- Logistics, transportation, and warehousing
- Purchasing procurement and sourcing
- Inventory and materials management
- Forecasting, planning, and scheduling
- Supply chain software and technology

The secondary functions included:

- Manufacturing
- Supplier relationship management
- Customer relationship management
- Marketing, sales, and customer service

We interpret these findings to indicate that, once again, most business organizations are hard at work on the basics, or level 1 through 3 efforts, while postponing higher level efforts that would include a closer network type of relationship with suppliers and customers. Virtually 90% of the firms responding to the survey placed their companies or business units in levels 1 through 3, while there was evidence that the ASCM concepts were not fully understood or applied. The only firms giving themselves multiple level 4 or 5 ratings were in the discrete manufacturing, high-technology, telecommunications, and wholesale distribution industries. The laggards formed a much larger group, with defense, government, oil and gas, publishing and printing, and utilities contained in that area. Best-in-class ratings were given to Dell, Wal-Mart, Procter & Gamble, Toyota, Cisco Systems, IBM, General Electric, and Hewlett-Packard.

The most surprising result of the survey was the indication that a solid supply chain strategy was generally missing in action. By that we mean most responding firms indicated they did not have a specific strategy at work to define the supply chain effort or, more importantly, to link the results with the business plan. One of the purposes of this text will be to demonstrate the importance of such a clear strategy and how a link to the business plan can result in substantiated improvements to sales and earnings, found directly on the profit and loss statement prepared by the firm.

In conclusion, the survey did document savings as a result of supply chain efforts. When asked what has been the overall impact of the supply chain effort on costs, those responding indicated:

- Reduced by 1 to 5%   27% of responses
- Reduced by 6 to 10%   33% of responses
- Reduced by 11 to 20%   8% of responses
- Reduced by more than 20% 4% of responses
- Initiative failed to meet objectives   6% of responses
- Not sure   21% of responses

When asked what has been the overall impact of the supply chain efforts on increased revenues, the answers were not as definitive. Responses indicated:

- Increased by 1 to 5%   27% of responses
- Increased by 6 to 10%   21% of responses
- Increased by 11 to 20%   5% of responses
- Increased by more than 20% 4% of responses
- Failed to meet objectives 5% of responses
- Not sure   38% of responses

We interpret the high percentage of responses indicating a failure or not sure as an indication that the idea of using supply chain to differentiate a firm and its network partners from competitors to create new sales is still not an integral part of supply chain efforts in general. Building top line performance becomes the future challenge, as companies discover how to use supply chain excellence in both directions on the profit and loss statement.

By combining the survey results with CSC research from other firms indicating a solid supply chain effort, Figure 1.2 can be used as a guide to just what might be achieved from a continued and concerted effort to use supply chain to improve both top (revenue) and bottom (cost) line results and how five to eight points of new profits are possible.

To use this chart effectively, a company should consider the bottom, horizontal axis as indicating the percent of profit made before paying taxes. The improvements are then measured in additional points of new profit as a percent of net revenues. Beginning on the left of the chart, there can be no doubt that inventory reduction is a viable part of a supply chain effort and can result in

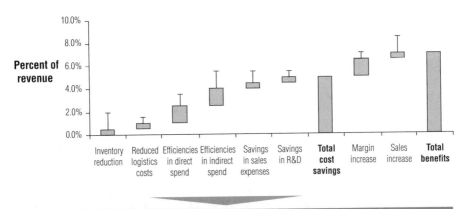

Figure 1.2. The Potential Savings and Revenue Enhancements: Projected Benefits from Collaborative Initiative Implementations (Source: CSC projections based on benchmarked performance of comparable solution offerings)

as much as a two-point improvement in profits. We indicate that a much smaller amount is actually achieved, because, as mentioned, most inventories are simply reallocated throughout the network, rather than eliminated. Inventory remains a frontier to be fully conquered, as supply chain efforts mature, and demand is more closely matched with supply, visibility into the supply chain network is established, and the need for extra stocks is eliminated. A firm using the chart is well advised to establish a realistic target for this part of the improvement effort.

Next we see an improvement through reduced logistics costs, resulting in a possible point of improvement in earnings. Once again, a range is used because most firms neglect to work as diligently on improving inbound freight as they do on internal handling and outbound freight. These companies also tend to overlook the possibility of using supply chain partners for storing and moving some of the goods or fail to apply some of the contemporary logistics concepts that include virtual ownership of assets and shared use of equipment and facilities. (For a more detailed consideration of the contemporary view of logistics, see Poirier, 2004.)

The areas of direct and indirect spending merit the largest potential benefits, as purchasing covers the largest portion of costs in most businesses. We find that the earliest efforts (levels 1 and 2) typically bring in a reasonable return and establish sufficient savings to validate the effort, as indicated by the large range of possible improvement. Later progress by most companies will focus attention on the indirect spend, moving into areas such as office supplies, furniture, travel, computers, office equipment, and so forth. Now there is a second level of savings possible, particularly as elements of e-procurement are introduced, reducing the buying and selling expense and drastically lowering the transaction cost.

Savings from sales expense are included, as there should be a lesser need for service personnel and special expenses to make heroic efforts to meet actual customer needs. An advanced-level supply chain gets deep into customer relationship management and equips the organization with the kind of information that requires far less time spent on responding to problems and much more on managing the relationship for mutual benefit. Savings in research and development are also included, as any advanced-level effort will include a collaborative effort to bring innovative products and services to market, as well as dramatically improve the cycle time from concept to commercial acceptance with a higher rate of final acceptance into the market. As indicated, the improvement in earnings as a percent of revenue can be as much as three to five points, resulting essentially from cost savings within the supply chain system.

Turning to the external environment, we find two more categories of improved benefits. First, there is the opportunity to increase margins by gaining higher selling prices for documented savings, resulting from demonstrated supply chain improvements. The growing number of case studies showing the potential to create sales lifts as a result of better collaborative sales events, promotions, and joint selling efforts supports the contention in the graph that another point of profit is possible through top line efforts.

In total, the chart indicates that a range of five to eight points of new profits is feasible. A firm should use this type of chart to establish the range of expected improvements from its efforts. Other categories of improvement can be included and different ranges indicated. The purpose is to have a document that establishes the expected end line for a concerted ASCM effort. When tied to a time line, it becomes a living document guiding the overall effort and establishes the fundamental basis for what should appear in future profit and loss statements.

IMPLICATIONS FOR THE CROSS-ORGANIZATIONAL NETWORK: A GENERIC ENTERPRISE BUSINESS PROCESS MODEL

The key question at this point becomes: How do supply chain processes create more value as the enterprise matures and expands? With determining how to improve business process at the forefront of management thinking, driven to reap the benefits illustrated in Figure 1.2, the opportunity becomes one of moving a firm through a logical progression to achieve the highest possible benefits. Along the way, the firm moving toward the intelligent value network wants to quantify the potential value of the transformation. That effort requires the creation of a process maturity roadmap, which follows the maturity model and leads the firm to becoming part of the best process-enabled extended enterprise and eventually an intelligent value network.

To create such a map, we need a process model of the extended enterprise, one that in particular illustrates the key supply chain processes that the enterprise executes. A process model that has become a standard in many industries over the last few years is the Supply-Chain Operations Reference (SCOR®) model, developed by the Supply-Chain Council, an independent, not-for-profit, global corporation with membership open to all companies and organizations interested in applying and advancing the state-of-the-art in supply chain management systems and practices. The SCOR® model captures the council's consensus view of supply chain management.

For the supply chain practitioner, SCOR® provides an excellent starting point for developing supply chain thinking, and we have adopted it to guide our thinking about process maturity. SCOR® covers the processes associated directly with the supply chain — its design, planning, and execution. In the authors' experience, the way in which a product is designed for manufacture has a marked effect on the subsequent effectiveness of the supply chain. The time to get a new product to market is often critical in maximizing the return from the product, and the way in which changes are managed can make or break a product's long-term profitability.

We have augmented SCOR® with a set of processes that impact the Make portion of the model and cover the engineering steps — processes that determine the product design, how it is to be manufactured, and how the design will be modified over time. These processes are illustrated in Figure 1.3, between the need to manage the supply chain and the need to manage the products manufactured. They are indicated as the linked arrows impacting the Make phase and include (a) the design phase that is impacted by manage product portfolio, design product, configure product (concurrently), and change product and (b) the production engineering phase that is affected by design process and tooling and maintain process. Getting these processes done correctly is as important as any of the other major processes contained in the SCOR® model.

The familiar top-level processes in the SCOR® model are illustrated in Figure 1.4. The five key processes are:

- Plan
- Source
- Make

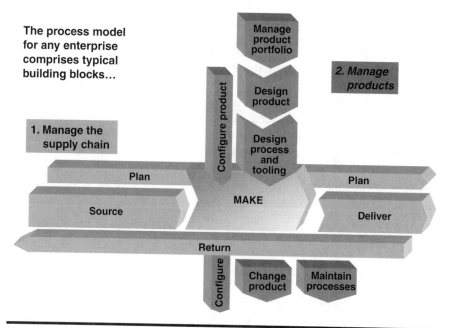

Figure 1.3. Process Maturity — Managing Supply Chain (Based on Supply-Chain Council SCOR® Model)

- Deliver
- Return

The *Plan* processes cover the work involved in identifying the requirements that the supply chain must fulfill and the resources at its disposal. These processes encompass the activities of matching those resources and capabilities to create a viable plan and the communication of that plan to everyone who needs to know. Plan breaks down into more detail, to represent the planning work for each of the next four process steps — Plan Source, Plan Make, Plan Deliver, and Plan Return. Also included are the policies that the company adopts which govern the way the supply chain will operate.

The *Source* processes follow a similar pattern. They cover the identification of suppliers, the communication of sourced requirements to those suppliers, and the creation and agreement of schedules for delivery. They follow the receipt of goods through verification and end with the sanctioning of payment to the supplier.

Make processes cover the detailed planning of production, the issue of materials to the production process, and monitoring of progress on the shop

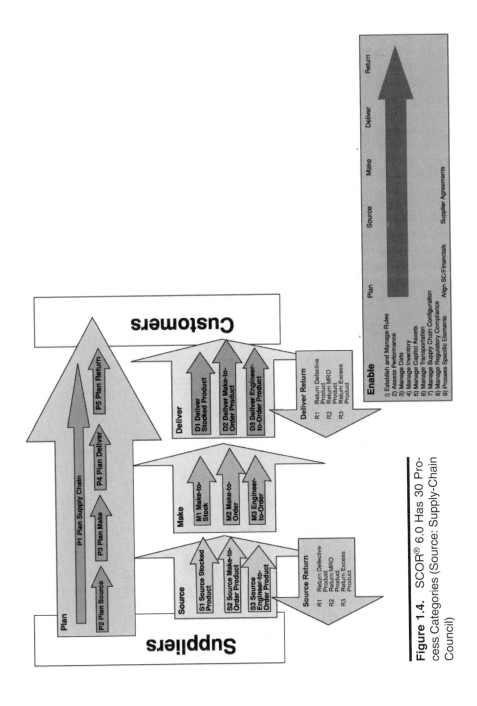

Figure 1.4. SCOR® 6.0 Has 30 Process Categories (Source: Supply-Chain Council)

floor through to making the completed, tested products available to the dispatch processes.

Deliver completes the traditional supply chain, getting the goods to the customer. It encompasses many of the "order-to-cash" processes, from inquiry and order processing through final product configuration, warehousing, order picking, and logistics planning (load building, routing, dispatch planning) to collecting proof of delivery or installation and invoicing. A version of Deliver deals with the delivery of retail products to the back room of the store, onto the shelves, and through the cash registers via the shopping cart.

Recent additions to the SCOR® model are the *Return* processes — a need driven by new legislation on recycling and packaging, along with the need to provide ever-higher levels of after-care service to customers, at an affordable cost. The process elements deal with authorizing the return of a product (warranty claim, replace or refund), organizing the receipt of the product back into the supply chain, and planning the most cost-effective way of dealing with the defective product (whether to repair or scrap). Of course, repaired products need to reenter the supply chain, to flow back in the customer direction, so the Return process ultimately links back to Plan, where this additional source of product can be taken into account.

The engineering processes that complete our enterprise model are comprised of:

- Manage product portfolio
- Design product
- Configure product
- Design (manufacturing) process and tooling
- Change product design
- Maintain (manufacturing) process

The *Manage product portfolio* process encompasses all the activities associated with deciding what products to provide through the supply chain. It includes the generation and testing of research ideas, proof of concept, and the management of linked groups of products through options and features.

The *Design product* process takes a chosen product idea through to a viable product design ready for release to manufacturing. The process covers not only a complete finished product, but also the parts and assemblies that go to make the finished product. The process includes concept design, the development of the design (including testing of materials, piece parts, and assemblies), through to a production design, released to manufacturing. Product designs are first conceived as ideas in the concept stage. The process model for concept design must be highly flexible and nondeterministic in nature, with design ideas moving between suppliers and team members in a free-flowing, synchronous manner. Once the design ideas and concepts have converged, design intent then moves to the design/redesign stage. Here, the number of unknowns has lessened to the point where a controlled work flow model can be implemented with more formal approval and release control.

Configure product governs the way in which products are configured to ensure maximum value of common and standardized components and the reuse of previous design ideas. Configuration management controls the content of the product from design through manufacture and throughout its service life. It includes the creation of bills of material, support for revision and version control, support for "where used" searches, and multiple bill of material views. The product data integrate with enterprise resource planning systems, and more advanced companies will have support for rules-driven configurations. Product configuration management does not stop once the product is released from manufacture — it continues as an essential part of the management of the product's entire life cycle.

The *Design process and tooling* processes take us from a design to a set of manufacturing instructions and the necessary tooling to support the manufacture. Traditionally, this was a "one-hit" process or at best a two-hit! For example, British automotive manufacturers could take up to six months to produce a new set of tooling to iron out production problems. A the same time, Japanese manufacturers had cut the time to six weeks, so prototyping became the accepted way of developing the production process. Thus, for more mature companies, this process interacts with the *Design* and *Configure* processes. Finally, the completed product design is ready for production release, where the production process model takes over.

Finally, we have two processes that deal with the changing product design: *Change product design* and *Maintain (manufacturing) process.* Change management is critical to both the longevity of the product's profitability and the continued effectiveness of the supply chain to deliver the product. While we have shown the *Change* process starting after the product is released for manufacture, in fact best practice dictates that change control is applied from the beginning of the design process. Version control and change effectivity dates are crucial to managing the evolution of the design both pre- and post-production. In parallel with changes to the product's design, manufacturing processes must be kept up to date, both in terms of their own effectiveness and to keep in line with changes to the design. Our final process — *Maintain process* — deals with this requirement by establishing processes that make certain the designs and products are still viable years into their life cycle.

THE ROUTE TO ADVANCED BUSINESS PERFORMANCE

As mentioned, the SCOR® model and its engineering process counterparts provide an excellent framework for analyzing how to improve an enterprise's end-to-end processing. We will apply this model, in conjunction with the supply chain maturity model and maturity matrices (to be introduced), to explain the path to the highest appropriate level of supply chain progress, keeping in mind that there are widely differing interpretations of these processes by companies at different levels of supply chain sophistication. The simple silo-based planning of a level 1 company, shared in an ad hoc way with neighboring internal departments, for example, compares to the complex, reactive planning of a level 4 or 5 network, with collaborative, interactive comparisons of schedules up and down key threads in the network. The naïve "replace and scrap" policy for returns in level 1 supply chains contrasts with the more sophisticated approaches of higher level organizations, which are planning the cost-effective reentry of refurbished products into the flow.

It is these different levels of application of the basic process models and the value that can be generated that the authors have come to understand. For a firm to generate value from the supply chain journey and become part of the process-enabled extended enterprise requires the creation of a guiding process maturity roadmap.

Table 1.1 is a matrix that is useful for establishing such a roadmap. The matrix takes the top-level SCOR® processes — Plan, Source, Make, Deliver, and Return — and sets out the attributes that can be expected for each process as a company moves through the five levels of supply chain maturity.

The attributes are designed as a stretch for each level, and as our recent surveys have shown, few companies have moved much beyond level 3. However, there is merit in establishing an absolute framework against which progress and value creation can be measured, to avoid the problems of traditional benchmarking, where the top-level, best-in-class, attained performance may still only be at level 3. In subsequent chapters, we will lay out the results of extensive research in a number of case histories, to demonstrate the value that companies have actually achieved in moving through the levels.

To support the growing process maturity, tools and technology are often deployed. Again, we have developed a guiding matrix that indicates the types of approach that typically support the processes in Table 1.1. This technology matrix is shown in Table 1.2.

With these two matrices as our guide, we can describe the way in which the supply chain processes and their supporting technologies develop — the roadmap to greater supply chain value. Companies that want to follow this roadmap are well advised to customize the matrices and the SCOR® map to fit the actual circumstances of their industry and market conditions.

THE PLAN PROCESSES

Plan processes at their simplest, level 1, are point solutions. Plans are created internally, with no integrated processes between different functions or plants within the company. Basic tools — spreadsheets and simple material requirements planning (MRP) or inventory control tools — support the process.

At level 2, the planning processes pick up and aggregate demand across the firm. Simple MRP planning tools are augmented with finite capacity planning capability — S&OP and later advanced planning and scheduling systems — which allows the firm to create reliable plans that can be adapted to ensure that more focused performance goals can be met. The planning now takes place across functional boundaries, so the firm moves toward internal supply chain excellence.

Between level 2 and level 3 is the wall — between my company and your company. This wall acts as a psychological barrier to improvement, but the more innovative firms break through, extending their processes into their suppliers and customers. Now they can pick up advanced demand forecasts from customers and give their suppliers early warning of changes in market conditions and cancelation or reduction of supply rates. This condition increasingly takes place collaboratively, with direct, electronic interchange of data between planning systems, governed by strict rules of engagement, such as service level agreements between collaborating partners in the extended value chain.

Level 4 planning takes in more steps in the extended chain. Some firms can travel back four or five levels in the chains supplying key components. Dell, for example, can reschedule its supply base for key commodities across five levels of suppliers in 30 to 45 minutes. Now the whole process is enabled by technology. The various players in the extended value chain are linked together, across the Internet, with common data standards and linked process flows. Critical events are monitored by real-time software and alerts sent to control points in the chain if plans are not achieved.

No company or network, to the authors' knowledge, has yet fully achieved the aspiration of level 5. The logical conclusion of the integration of Planning across a supply network is to manage the network as though it were a single entity, with no barriers caused by the artificial accident of share ownership. Moving between level 4 and level 5 is similar to the internal supply chain, moving between level 1 and level 2.

Table 1.1.   Process Maturity: Managing Supply Chain

		Maturity Level 1	Maturity Level 2
Manage Supply Chain	**Plan**	Demand/supply planning is done internally, with no integrated processes and tools across plants.	Global demand/supply planning is consistently aggregated across the firm, focused on functional accountability, and continuously improved by comparisons to historical performance.
	Source	SRM partnerships are poorly defined. Processes are informal. There is no integrated set of tools to allow common access to procurement data.	Cross-functional commodity management teams and SRM partnerships are in place. Common ERP systems are used effectively.
	Make	Manual material and production control activities are driven by rudimentary implementation of MRP/MPS tools.	Material and production control data are tracked electronically to optimize internal scheduling and inventory management.
	Deliver	No formal, standardized processes or tools are in place for order management, channel rules, product delivery, or invoicing.	Formal outbound logistics processes, automated order management systems, specific channel rules (terms and conditions) and delivery standards, and automatic invoicing exist; variability in order entry and scheduling across product divisions.
	Return	Unplanned response to returns. Returned goods not netted off against requirements for reuse.	Planned response to returns. Not coordinated with upstream supplier. No visibility from customers of impending returns.

ERP = enterprise resource planning, MPS = master production schedule, SRM = supplier relationship management

Maturity Level 3	Maturity Level 4	Maturity Level 5
Strategic partnerships with customers and suppliers are facilitated by direct, collaborative, electronic data exchange and governed by formal supply chain performance agreements.	Dynamic global demand forecasting and capacity utilization calculations feed demand/supply decision-making mechanisms. Joint demand/supply decision-making bodies leverage and share data globally.	Planning is carried out for virtual enterprises as though they were single entities. Plan optimization based on value creation in the whole chain.
Strategic commodity/supplier partners participate in collaborative product development, process/TCO improvement programs, and consortia buying and have access to select on-line data.	Integrated supply network uses e-enabled systems to automate/ optimize all commodity and supplier transactions.	Value chains co-manage sourcing strategy and decisions, to ensure best value for all the players in the chain.
Customer-driven APS (linked to suppliers), kanban demand pull manufacturing, real-time inventory control, automated product quality control, and total life cycle product data management are dominant.	Fully enabled, electronically captured APS, product configuration specification, demand pull, inventory backflushing, product history, and quality control systems allow instantaneous product changes and drive continuous improvement.	Value chains co-manage longer term capacity investment decisions and short-term scheduling value trade-offs as though they were a single entity.
Product and delivery process data maintenance systems function simultaneously throughout the supply chain and are accurate and visible to all supply chain partners via e-commerce systems. Differentiated service levels and performance agreements are formalized.	Comprehensive e-commerce linkages throughout supply chain optimize warehousing (outsourced but integrated), tracking, transportation and delivery, and automated invoicing. Differentiated channel rules and order/service levels, including real-time commitments.	Logistics capabilities are shared between value chains, to maximize value creation between and across enterprises.
Visibility of customers returning goods. Planned response and handoff where appropriate to supplier.	Returned goods flow back through the supply chain seamlessly. Response policies are preplanned to maximize value.	Full visibility of returning goods to the whole network. Reuse, rework, or scrap policies applied automatically to maximize overall value.

APS = advanced planning and scheduling, TCO = total cost of ownership

Table 1.2.   Technology Maturity: Managing Supply Chain

		Maturity Level 1	Maturity Level 2
Manage Supply Chain	Plan	■ MRP: order management, inventory management ■ Point solutions, local silos	■ MRP II, DRP, APS ■ Internal connectivity ■ Enterprise intranet
	Source	■ Business unit aggregation tools	■ Enterprise aggregation tools
	Make	■ Rudimentary implementation of MRP/MPS tools	■ Shop floor information tracking fed back into MRP systems
	Deliver	■ Manual dispatch planning ■ Ad hoc links between order processing, picking, and dispatch planning	■ Warehouse management systems ■ Transportation management systems ■ Third-party logistics
	Return	■ Basic service requirements monitored	■ Some analysis of returns to establish improvement opportunities

APS = advanced planning and scheduling, DRP = distribution resources planning, MPS = master production schedule

Maturity Level 3	Maturity Level 4	Maturity Level 5
■ ERP, CPFR, capable to manufacture, available to promise, capable to promise ■ Linked to intranets, corporate strategy, and architecture	■ Collaborative network planning ■ Glass pipeline ■ Dynamic tactical planning, with decision support systems ■ Best asset utilization ■ Event management ■ Internet-based extranet, shared capabilities	■ Full network optimization ■ Shared processes and systems ■ Simulation techniques ■ Full network communication system ■ Shared architecture and planning
■ Full network aggregation tools ■ Supplier integration and synchronization	■ Web-based sourcing ■ e-procurement ■ Inbound logistics managed in same model as outbound	■ Network sourcing through best constituent
■ Advanced supply chain planning tools driven by customer demand ■ Local shop floor control by kanbans ■ Real-time collection of progress information	■ Full scheduling collaboration with immediate supply chain partners ■ Virtual manufacturing	■ Value chains co-manage their operations in shared scheduling models ■ Value trade-offs are made as though the network were a single entity
■ Pull systems — information from immediate customer ■ Vendor-managed inventory ■ Network planning ■ Information allows choice of best internal or external provider — lead logistics provider	■ Information allows choice of best constituent provider ■ e-fulfillment ■ Multitier logistics model	■ Global satellite positioning ■ Virtual logistics optimization based on total logistics cost
■ Returns managed in same logistics model as outbound ■ Preplanned responses ■ Data used as a source of improvement opportunities	■ Returns managed as a network opportunity ■ Focus moves to regaining lost values for all participants	■ Network allies assume responsibility for some parts of return process ■ Salvage, refurbishing, and reshipment become viable techniques

CPFR = collaborative planning, forecasting, and replenishment, ERP = enterprise resource planning

Between level 1 and level 2, performance improvements come from moving out of functional silos to internal supply chains, where performance is measured across the whole chain from inbound dock to dispatch. All the local functional targets — for purchasing, manufacturing, and sales — are subservient to achieving the best overall performance. The companies in the level 4 supply chain are the equivalent of the functional silos at level 1. Local performance measures are represented by local shareholder interests. To move to level 5, and realize the major value opportunities, the linked firms must work on optimizing the whole network, not individual shareholder funds.

SOURCE, MAKE, DELIVER, AND RETURN

Source processes follow a similar path to *Plan*. At level 1, there are few formal processes and relationships with suppliers are ad hoc. There is little ability to aggregate data across the firm, so sourcing decisions are suboptimal. At level 2, we find tools that allow the firm to build up a consolidated view of total demand — the best deals can be made, but from a traditional buyer-seller relationship. Across the wall to level 3, we see the development of well-defined supplier partnerships, with shared goals for driving and sharing extra value. Suppliers are integrated into the planning process and are probably involved much earlier in the design process, for noncommodity items. By level 4, firms will have adopted e-sourcing tools for commodity items, allowing them to search a much wider Internet community for the best supply opportunities. More and more of the procurement processes will be automated and linked into the scheduling systems on both sides and to back-office reconciliation and payment systems. The level 5 aspiration is for the whole network to co-manage sourcing strategy to maximize the value to all the players in the chain.

The *Make* processes at level 1 are often as rudimentary as the planning processes that feed them their instructions. Paper-based or simple MRP-based tools track the manufacture of products, relying on progress chasing to maintain due date performance. At level 2, tracking is electronic, and linked into the scheduling system, so that the whole internal enterprise has a view of the state of play. Customer delivery performance is greatly enhanced, and lead times and shop floor inventories are reduced. The level 3 firm takes its pace from direct customer information — real-time progress information can be shared with customers, so that problems are highlighted before they become critical. Simplified shop floor control through demand-pull, kanban-type systems is prevalent. When level 4 is reached, the collaboration with immediate supply chain partners is complete. By level 5, we are looking to make long-term, strategic capacity and technology investment decisions in the best interests of the whole value network — the activity of the whole chain is scheduled as part of a single model.

Deliver completes the customer-focused processes. At level 1, we again see manual planning, often with only ad hoc links to the preceding functions in the internal supply chain. The level 2 firm uses effective warehouse and fleet management systems to underpin good outbound logistics. By level 3, we see the start of information flowing directly from the customer to the supplier. Vendor-managed inventory reduces the assets on the customer's balance sheet and allows the supplier to use superior knowledge to reduce the total cost of supply. Better information facilitates the choice of the best service logistics provider. The level 4 company plays in an extended value chain, so there are more opportunities to collaborate in multitier logistics systems, to optimize warehousing and transportation costs. Network companies recognize that distribution is a noncore activity and collaborate to reduce their combined logistics bill. Aspirations of the level 5 company, in its total value network, encompass Web-based virtual logistics operations, where the service provider can track goods in real time using GPS technology, to maximize customer service, while minimizing the total cost.

The last set of processes covers *Return* of goods, faulty or in need of refurbishment. This is a fairly recent development. Historically, returns have been handled in an ad hoc manner. The firm's response depends on who takes the call. Many level 1 firms still adopt this approach. By level 3, firms have visibility of returning goods, so that effective response can be preplanned and possibly handed off to a supplier. A level 5 firm would expect to have access to full visibility of the return flow of goods in the whole network — and, of course, the number of returns would be minimized!

AUGMENTED SCOR®: THE ENGINEERING PROCESSES

Design and production engineering go hand in hand with supply chain improvements efforts. Ineffective product design or design for manufacture and supply inhibit profitability just as much as poor supply chain design and execution. In fact, the two sets of processes are complementary, and we see the same five levels of maturity — firms at level 1 in their supply chain thinking tend to be level 1 in their design and production engineering thinking.

The maturity matrices for the design and production engineering processes are shown in Tables 1.3 and 1.4. The matching technology maturity matrix is provided in Table 1.5.

Table 1.3.   Process Maturity: Design Engineering

	Maturity Level 1	Maturity Level 2
Manage Product Portfolio	■ Users are primarily from engineering ■ PDM is used in one or a few key functions (e.g., MCAD, ECO) ■ Product structures are limited to "stand-alone" bill structures	■ Basic PDM system integrated with primary authoring tools (business and technical) and downstream ERP system(s) ■ Product portfolios are managed via traditional spreadsheets
Design Product	■ Product development groups are typically co-located and working independently, with late partner and supplier support ■ Product service is considered up front but is supported by separate databases ■ Product data "push" to suppliers and customers ■ Collaboration is performed through a mix of media and mostly not in real time	■ Expansion of formal product data creation and management through the enterprise via PDM/PLM ■ Ability to view common formats (e.g., pdf) across the enterprise ■ Access control is granular — by user/group, data type, project, etc. ■ Access to PDM provided to selected partners and suppliers (e.g., for sharing of CAD/CAM files with contract manufacturers, pattern makers, etc.)
Configure Product	■ Data management relies on legacy systems with incomplete functionality (vaulting, product structure, work flow, visualization) ■ Product information creation tools are not integrated with the data management system(s)	■ PDM assists in rationalizing components and sourcing
Change Product	■ Change processes focus on engineering and are largely manual	■ Projects are coordinated among domestic or global locations using traditional communication (telephone, fax, e-mail, and face-to-face meetings)

ECO = engineering change order, ERP = enterprise resource planning, MCAD = manufacturing computer-aided design, PDM = product data management, PLM = product life cycle management

Maturity Level 3	Maturity Level 4	Maturity Level 5
■ Use of a comprehensive PDM for shared vaulting, product structure, work flow, visualization ■ PLM supports base + option (variant) bills, alternates, substitutes, multiple views	■ Use of PLM accessible via Web connections to manage/view information through entire product life cycle ■ PLM assures broad distribution of product information via notification and subscription services ■ PLM supports product portfolio analysis and resource management (with an internal scope)	■ PLM systems are connected (federated) among key firms ■ Develop joint input on product direction with customers and suppliers (tiers one to three) ■ PLM supports product portfolio analysis and resource management (including the external scope)
■ PLM access given to some extended design partners and suppliers on the NPD/NPI team ■ Tasks are allocated among several global locations using PLM	■ Access to PLM provided to third-party design partners/suppliers on the NPD/NPI team ■ Teams utilize concurrent engineering — including design for supply chain ■ PLM helps compress time to commercialization and scale ■ Leverage the technology of key suppliers across portfolio of existing and future products ■ Capture and share all design rationale with business partners ■ Collaborative real-time design with second- and third-tier suppliers ■ Design for CTO/ETO (looking toward mass customization)	■ Access control, distributed work flows (to third parties) and browser-based viewing ■ Empower strategic suppliers across all aspects of the product life cycle (concept through end of line) ■ Product development, using leading-edge PLM tools, is structured to happen anywhere, be coordinated from anywhere, to be manufactured anywhere
■ PLM supports reducing number of component suppliers	■ Product development using PLM is structured to happen anywhere and be coordinated from anywhere ■ Product support can be via third parties that access PLM	■ PLM supports key enablers of mass customization of products
■ PLM supports enterprise change processes serving all functions (engineering, manufacturing, suppliers, service) ■ Project management is integrated with PLM work flow and cost reporting systems	■ Design changes managed using PLM are structured to happen anywhere and be coordinated from anywhere	■ PLM directly supports value-added services into the product distribution process (e.g., postponement of some final work)

CTO = configure to order, ETO = engineer to order, NPD = new product development, NPI = new product introduction

Table 1.4. Process Maturity: Production Engineering

	Maturity Level 1	Maturity Level 2
Design Process and Tooling	■ Manufacturing processes established after design complete ■ Products have long quality maturation and many design concessions	■ PDM is used by concurrent engineering teams for fabrication, assembly, test, service ■ PDM assists in rationalizing components and sourcing
Maintain Process	■ Process knowledge with the operators ■ No standardization of processes	■ Standardized processes, documented ■ Other processes ad hoc

PDM = product data management

At level 1, the *Manage product portfolio* process is often the preserve of just the engineering department. As maturity increases to level 2, other departments become actively involved in the process — sharing ideas and decisions in a way that enhances the whole enterprise. Suppliers and partners get in on the act at level 3. By level 4, there is evidence of shared thinking about product development up and down the extended value chain.

A level 1 company executes the *Design product* process in silos, just as it runs its internal supply chain. Products are invented by marketing, designed by engineering, and made by manufacturing. By level 2, there is integration across the whole internal process — an organized set of interactions to ensure that a product is designed in a way that maximizes the company's total profitability. In level 3, the suppliers of key components and assemblies are also becoming involved in the design process, which extends, in levels 4 and 5, to ever-greater collaboration, supported by Web-based tools for product data management and computer-aided design.

The *Configure product* process can create great problems at level 1, as it is, by its nature, an integrating process. With the design process fragmented, configuration management is difficult. The problem is solved within the company at level 2, using product data management applications to manage the product structure, record new versions/revisions, and plan effectivity. As suppliers become involved, at level 3, the process often causes the number of suppliers to be reduced, as both sides invest in the growing relationships. By

Maturity Level 3	Maturity Level 4	Maturity Level 5
■ Tool integrations assure capture by manufacturing of associated information (packaging, pubs, instructions) ■ PLM supports BTO, CTO, ATO ■ Rich PLM relationships tie parts to documents, process plans, etc.	■ Design for CTO/ETO (looking forward to mass customization)	■ PLM directly supports value-added services into the product distribution process (e.g., postponement of some final work)
■ PLM directly supports product service (help desk, specifications, repair parts, reference drawings, etc.)	■ All processes are fully documented and linked to the PLM	■ Simple, mature processes that may be executed anywhere ■ Process changes shared across federated enterprise

ATO = available to order, BTO = build to order, CTO = configure to order, ETO = engineer to order, PLM = product life cycle management

level 4 and 5, the collaboration reaches a stage at which configuration is managed across the extended chain and can be coordinated from anywhere.

At level 1, tooling and manufacturing processes are established after the product design is complete. By level 2, the *Design (manufacturing) process and tooling* process works in tandem with *Design product*. A level 3 company has integrated the processing to the stage where the manufacturing process can support more rapid changes — configure or build to order, incorporating customer order requirements directly into the manufacturing and assembly processes. By level 5, the network has a full mass-customization capability, with postponement of final work right into the logistics chain.

The final two processes — *Change product design* and *Maintain (manufacturing) process* — both follow similar paths. The level 1 company lives within its functions. The level 5 company plays in networks, collaborating with its partners to maximize the value they can jointly create.

These grids should all be customized to meet the needs of the individual firm and its business network allies. They should accompany any analysis and development plans as the firm moves to its highest level of appropriate maturity in the SCOR® and supply chain model. They will guide our own consideration of how a company and its trusted allies move toward supply chain management, and use ASCM in conjunction with business process management tools, to be described in the next chapter, to attain a significant advantage over less capable networks.

Table 1.5. Technology Maturity: Design and Production Engineering

		Maturity Level 1	Maturity Level 2
Design	Manage Product Portfolio	■ Continue to produce existing lines, SKUs, product varieties	■ Internal-only focus on any extensions or modifications to the portfolio
	Design Product	■ Little change or new introduction ■ General following or copying of successful products	■ Project management is traditional — project management tasks not updated (in sync) with PDM work flow results ■ Projects are coordinated among domestic or global locations using traditional communication (telephone, fax, e-mail, and face-to-face meetings)
	Configure Product	■ Localized configuration based on internal engineering and design specifications	■ Drive for standardization and use of viable common components ■ Building of design database for potential sharing with external parties
	Change Product Design	■ Little analysis of need for change ■ Wait until it fails	■ Selective analysis of mature products ■ Reaction to recalls
Production Engineering	Design Process and Tooling	■ Manufacturing instructions for use of basic tools and traditional processes	■ Some introduction of techniques to match best processing with available tools
	Maintain Process	■ Wait for drastic product decline before scheduling maintenance	■ Some planned maintenance/preventive techniques

PDM = product data management

Maturity Level 3	Maturity Level 4	Maturity Level 5
■ Selective use of external advice to impact portfolio ■ Cautious analysis of SKUs or total product availability ■ Some purging of nature products	■ Serious evaluation of portfolio with determination to purge items beyond useful life	■ Collaborative efforts to design portfolio that matches trends and demands from mutual database analysis
■ Specialized integrated tool sets (e.g., software, electrical, and test engineering) are integrated with PDM ■ Rich PLM relationships tie parts to documents, process plans, etc. ■ Project management is integrated with PLM work flow and cost reporting systems	■ PLM supports use of cross-enterprise scheduling tools to keep distributed projects in sync ■ PLM supports direct linkage of work flow and cost systems with partners and suppliers	■ PLM is an enterprise activity ■ Linked businesses analysis provides best response to assure high "hit" rates ■ Centralized work flow processes in effect
■ PLM directly supports product service (help desk, specifications, repair parts, reference drawings, etc.) ■ Cross-enterprise work introduced to improve configurations	■ Revisions, version control based on collaboration with key suppliers and customers ■ Integration across enterprise resource planning systems	■ Rules-based configurations based on life cycle attributes as analyzed by network partners
■ Systematic analysis of need for change, based on reliable metrics	■ Challenges to a portion of product line on a planned basis ■ Mutual studies of performance	■ Need for change linked to need to purge ■ Products kept viable by network
■ Joint analysis of best partner practices across emerging network	■ Collaborative assessment of best processing and most effective tooling	■ Jointly developed process models drive tooling and new designs
■ Specific manufacturing maintenance to sustain hierarchy of processes	■ Network collaboration on maintaining linked processes	■ Electronic procedures to systematically improve most viable manufacturing processes

PLM = product life cycle management

SUMMARY

The route to supply chain effectiveness is a journey, not necessarily an easy one, across five levels of maturity. Companies that embrace the challenges and reach the advanced levels are opening a significant gap in business performance between them and their competitors. Information in support of the positive results has been documented, indicating the possibility to add five to eight points of new profits to a firm's earnings. This same reporting also indicates, however, that most firms tend to bog down in midstream. The use of the maturity model, the SCOR® framework, and a set of maturity matrices provides a reference to track progress and understand what further improvements lie around the corner. These models and matrices answer questions such as:

- "How well are we doing?"
- "Have we covered all the bases?"
- "How much more value could we generate, and how do we get there?"

A firm interested in achieving the highest level of pertinent progress is advised to calibrate its progress on the maturity/SCOR® models and develop a set of matrices pertinent to its industry, market, and environmental needs. With these tools, the company can begin to selectively collaborate with business allies and pursue a sensible path to the future. Using the SCOR® model as a backdrop for determining how to improve the processes involved can also match the effort with a generally accepted methodology for finding beneficial changes.

In the next chapter, we extend the use of the model and matrices, as we describe some of the emerging technologies that today make the process world so much more accessible to the forward-looking business.

2

BUSINESS PROCESS MANAGEMENT AS THE ENABLING INGREDIENT

To transition a business into the higher levels of the supply chain evolution, and into the realm of advanced supply chain management (ASCM), firms must accept two necessities as they overcome the typical barriers created between levels 2 and 3 of the maturity model. First, there is the general unwillingness to share valuable information outside of the four walls representing the internal organization. This unwillingness is first seen in level 1 companies, between departments and business units, and becomes one of the first barriers to overcome. Most companies welcome useful insights from external constituents, but stalwartly resist sharing information they believe gives them an edge in an industry or market. The solution here is to establish what can or cannot be shared through collaborative efforts early in the discussion and then to demonstrate the value of the sharing in terms of increased, documented benefits for both parties. Typically, some form of pilot operation or sharing effort is necessary with a particular business unit and selected external allies, to establish those parameters.

Second, there is the need to link many legacy systems and disparate software together in a manner that facilitates rapid exchange of the knowledge considered pertinent to the extended enterprise processing. That means the parties to the collaboration effort must decide exactly how to connect the constituents into a meaningful communication network without being stopped by difficulties with the age or type of database system holding the needed information. In this case, the secret is to establish a simple methodology for accessing portions or components of the different databases and extracting information that facilitates the creation of a better network response to actual market or customer needs. Business process management (BPM) becomes the key to success in both areas, as it introduces the possibility of easily sharing pertinent knowledge that helps all supply chain constituents and overcoming the problem of dealing with many different systems. This sharing can be done while protecting the security of the vital information, on a real-time basis across organizations of virtually any size in any business, building future viability in a collaborative manner.

Before we continue with our discussion of how a company and its business allies can make the most progress with the supply chain maturity model in conjunction with the SCOR® format, we need to establish the value of BPM in facilitating the required reengineering of the process steps that makes viable progress feasible. In this chapter, we consider the growing complexity of business networks and the value opportunities presented through the ability to easily enter into disparate information systems and extract the vital knowledge so necessary for higher level business interactions and attainment of level 3 and above positions. The basic concepts surrounding BPM were presented in an earlier book, *The Networked Supply Chain* (Poirier et al., 2004), and we offer that text as reference. In this book, we will summarize the important tenets and go further to cover the new technologies that form the backbone of BPM systems and explain how they open up opportunities for business users to take back ownership of supply chain processing and reap the greatest benefits from ASCM.

We will explain that the impediment to business progress in this area is not people but technology, and the ability to continuously reengineer critical processes in real time, without major information technology (IT) rework, gives the users a significant advantage. These new approaches have an entry fee, however. They require a fundamental rethinking of the role of the supporting IT architecture and driving systems. The good news is that the changes bring business benefits in the form of reduced costs and greater flexibility in the delivery systems — a real win-win condition. To begin this explanation, we must understand the complexity being faced in today's business environment.

KNOWLEDGE TRANSFER IS EXTENDED ACROSS COMPLEX BUSINESS NETWORKS

A central theme that has begun to pervade the supply chain agenda, virtually on a global basis and across nearly all industries and business environments, is that ASCM efforts eventually result in a demand-driven supply network and finally in what we term the intelligent value network. With that progression comes some very specific needs. To satisfy the inputs from a demand-driven business system, the linked enterprise businesses must have the ability within their network to accept reliable demand information, from a variety of sources, and share it quickly and accurately. Further, the resulting actions and work flows created by analysis and response to that knowledge must be synchronized and visible, so the parties involved have a common view of what is taking place. The result is an accurate and effective linking of the supply responses to the actual customer or consumer demands, without the necessity for excessive safety stocks and inventories.

As we develop our position on how the best ASCM efforts work diligently to create the best response system and thereby better satisfy customers and consumers, we will describe how the processing can be developed to meet these requirements. We must emphasize before that discussion, however, the need for timely and accurate knowledge transfer across what is becoming a very complex business environment. John Fontanella and others from AMR Research framed the situation well, when they stated: "Regardless of whether demand signals are forecasts, actual consumption data, or the result of collaborative planning between supplier and customer, mobilizing the supply chain to respond quickly and efficiently is the major challenge that companies face over the next five years" (Fontanella et al., 2004, p. 1).

One central purpose of any advanced supply chain effort is to create the kind of connectivity across the extended enterprise that results in end-to-end visibility into what is happening. Another is to establish a system that uses the latest technology to transfer data and ultimately valuable knowledge from the processing, so the network can mobilize and respond better than any competing system to the actual needs of the marketplace and key customers. A third is to be in a position of either introducing or always being near the front of the inevitable product or service designs and innovations that will enter the market. Satisfying these purposes occurs when the firm and its business allies meet the opportunity by linking together five topics of importance: ASCM, customer relationship management, supplier relationship management, appropriate technology applications, and customer intelligence, all factors that can be positively affected by BPM. The last topic is our terminology for the acquisition, management, and integration of customer knowledge in order to create a differentiating customer value proposition. As we move forward with our considerations, we will elaborate on the role of each factor.

To begin, if a firm looks holistically at the five, usually disparate, factors mentioned and determines how a leading position can be gained through supply chain excellence, it can develop integrated strategies and solutions for delivering products and services to key customers better than any competitors. When the effort is extended through BPM techniques, to include willing and trusted business allies, working across an extended enterprise for the same purposes, the advantages are unmatched. In theory at least, the collaboration that will occur across such a holistic effort will move the linked business network toward conditions of total enterprise optimization. In such an environment, the involved firms will be making the greatest use of assets, will be at or near lowest cost position, and will be generating the most new revenues by virtue of greater overall satisfaction to the intended customer or consumer groups at the end of the intelligent value network. In other words, they will have created a competitive advantage.

Any discussion on the possibilities of achieving such advanced supply chain management conditions and something like total enterprise optimization must begin with an understanding of just how complex an extended enterprise has become and why there is such a strong need for cross-organizational communication. The original supply chain efforts were directed toward achieving optimum operating conditions across a linear set of tightly linked, internal process steps — from beginning raw materials to final delivery of products and services. As the diagram presented in Figure 2.1 shows, most advanced supply

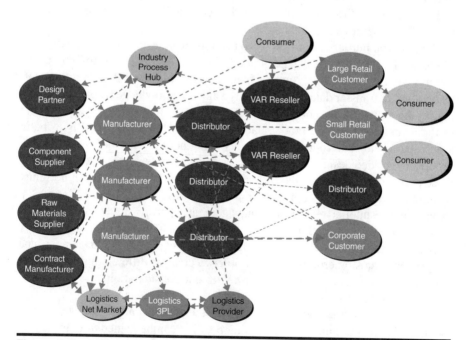

Figure 2.1.   Supply Chains Are Becoming Collaborative Networks

chains are becoming complex business systems, or what we term collaborative networks. There are so many business constituents to any ASCM extended enterprise that simply keeping track of them can become a challenge.

On the upstream side, there can be multiple supply partners, and not just raw material suppliers, including those helping with new product design or product offerings, component suppliers, or contract manufacturers. In the middle, there could be linked manufacturers, each making part of the final product, such as Boeing and its major manufacturing partners engaged in constructing the new 7E7 airplane. Contractors in Japan, Europe, and the United States will be involved in that endeavor. Distributors often play a key role in moving the products to market as well, and because of their ability to service remote geographies, multiple constituents might be used. In high-technology equipment, it is usual to have value-added resellers involved in the processing. At the top of Figure 2.1, we note that industry process hubs could be used, as well as industry marketplaces or some form of portal for handling part of the processing, particularly the sourcing function. At the bottom, we note several constituents handling part of the logistics function. When we then consider that there are now many types of end customers and consumers, the picture does indeed become complex.

Any supply chain analysis that is limited to internal processing, which is a small segment of the inter-enterprise business system, is doomed to operate with suboptimized conditions, because the focus is only on a part of the total network illustrated in Figure 2.1. There are simply too many players in a typical business network, creating the possibility that one or more can introduce complications that negate a part or most of the savings generated internally. The end-to-end processing that has come under scrutiny for improvement by the industry leaders now includes a multitude of business partners. Concurrently, the necessary flow of information and knowledge within a business network has become as important as the physical flow of goods and the transfer of money across what is clearly an extended enterprise. Without access to that information, a system is doomed to fall short of the intended optimized conditions.

The next issue deals with that need for information, which is often supplied through the use of the appropriate technology and systems. As indicated in Figure 2.2, we see that progressing through the supply chain maturity model requires matching enabling technology with the process changes occurring between levels. In the early levels (1 and 2), we see that the focus on divisional or business unit performance instead of the total organization drives firms to seek help with process optimization. Here we note the introduction of transportation management systems and warehouse management systems, along with efforts to improve order management and inventory management. Most com-

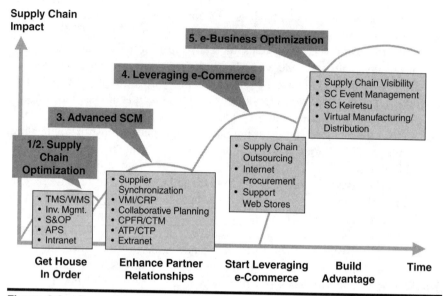

Figure 2.2. Progressing Through the Supply Chain Levels

panies also use the SCOR® model to help start the processes with a form of production scheduling based on existing forecasting techniques.

In the third level, where the firm continues its focus on attaining internal excellence, sales and operations planning appears. Supply chain optimization now requires the collaboration of a host of business partners working across the whole firm in concert for the same end results, a key element of which is generating reliable demand signals — to be matched with appropriate supply capabilities. To progress beyond this level, it becomes imperative in such an environment that the firm seeking optimized conditions make a passage from an internal-only perspective, in terms of generating selfish process improvements, to one where willing and trusted business allies are made a part of the process improvement effort, with the end result focused on customer satisfaction. To accomplish this objective, the leading firms are merging their ASCM concepts with those of a small number of trusted business allies to establish improved operating performance. The typical firm might not have all the constituents depicted in Figure 2.1, but most have a significant list of important companies that directly impact overall performance. The ability to extract and share important business intelligence becomes the value to be derived, for both the top and bottom lines of the profit and loss statement.

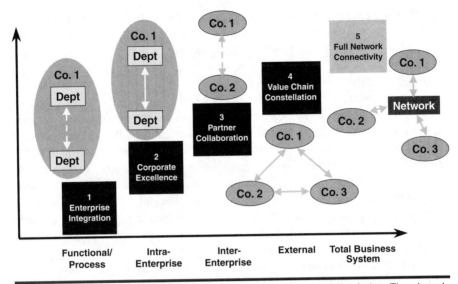

Figure 2.3. Organizational Barriers: Moving Between Each of the Five Levels Involves Overcoming Difficult Barriers to the Desired Changes

With this understanding, the firm must first attack the internal resistance to the kind of information sharing being advocated. Referring to Figure 2.3, we see the difficulty of moving to ASCM levels. In the lower levels, the problem is achieving department-to-department or business-unit-to-business-unit cooperation, not an easy task for many large corporations used to fostering strong internal independence. Next there is the issue in level 3 of beginning cross-company collaboration and serious sharing of vital information. As the network begins to form in level 4, the need is to include multiple businesses in the network, and the complexity increases as the value chain evolves. Finally, in the highest level, the companies work together in a collaborative manner, gaining access to knowledge across the total network connectivity. The transition described by the figure is not an easy one to accomplish and generally takes years of effort.

It starts with a concerted effort to link up the internal business units and functions, so the total leverage of the organization can be brought to bear on supply chain opportunities. It continues cautiously with the help of a few trusted external business allies. Bear in mind that very few businesses have reached anything close to level 5. The survey conducted by *Supply Chain Management Review* and CSC verified that most businesses are bogged down in levels 2 and

3, and virtually no respondent indicated his or her firm had achieved the highest level. We realize that every firm does not need to reach that level, but all should know what level in the progression it is appropriate to attain. For these reasons, the firm is advised to again determine how far in the progression it is willing to go, based on the expected benefits.

As progress is made, and the external environment proves not to be as threatening or insecure as expected, a variety of new opportunities appear. Figure 2.4 presents a description of some of those features. With a collaborative network established and the necessary communication system working to transfer vital information, the linked organizations can now begin to share their efforts in earnest. New materials can be considered for existing or new products. Cooperative development can be done with the help of willing allies and not be restricted to the internal R&D department. New processing can be considered in great detail, with the emphasis on making sure the most able partner is performing the key process steps.

Product adaptations and integrated solutions appear as people who never had a chance to interface with suppliers and customers are given the opportunity to work together to draw out the absolute best practices. Process improvements invariably result, as well as new products that reflect a variety of inputs and considerations. Supply chain responsiveness becomes a network responsibility, and the business allies work on matching supply with actual demand and providing the customers with unmatched service. Alternative channels are considered to make certain the right way is used to deliver the desired products and services.

From a very important aspect, the partners work together on the cost to serve, using activity-based costing or balanced scorecard techniques. The idea is to get a solid handle on exactly what the costs across the end-to-end supply chain are and how close to optimized conditions can be achieved. Done well, this part of the effort results in documentation of the benefits for all participants, as we will describe in subsequent chapters.

Now the effort moves from a bottom line orientation to how the network uses its newfound capabilities to increase revenues. The linked companies work together to find new and profitable sales, often making joint calls and presentations to existing and prospective customers. Sales leadership strategies are worked out in concert with key suppliers and major customers, so that the resulting new business systems carefully match what the end customers and consumers deem important with what the network can provide better than any competing system.

Across the bottom of the figure, we see how the major elements of the supply chain are enhanced as product development, order fulfillment, distribution efficiency, marketing strategy, and product replenishment reach new levels

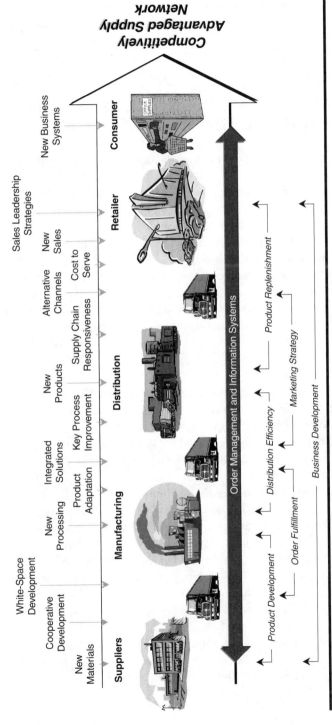

Figure 2.4. Advanced Supply Chain Management: Building Network Business

of efficiency by virtue of the combined efforts at work. Together these elements emerge as a new business development effort, based on an enhanced business model that takes advantage of the maturity model, the SCOR® framework, and the matrices in Chapter 1 to achieve the appropriate highest level of progression.

SUPPLY CHAIN PROGRESSION REQUIRES A SUPPORTING ARCHITECTURE AND ENABLING TECHNOLOGY

As companies move along the maturity model with helpful business allies, they reach the point where decisions need to be made regarding connectivity across the extended network. No firm has all of the knowledge necessary to reach optimized conditions. In spite of the very impressive profits we have seen some firms return, we always find the opportunity for more, if we can only convince the players to try a bit of inter-enterprise data sharing. Through whatever system is selected, ASCM requires linking network business allies into a coherent system of knowledge exchange. As the devices used (computers, wireless networks, personal access equipment, and so forth) become more reliable and globally connected, the chance to share data increases greatly. The information exchanges, moreover, become more meaningful as they improve in reliability, speed, and cost. Customer satisfaction under these conditions becomes more than a rallying cry. It sets the new parameters, to be met by the intelligent value network.

Meeting these challenges requires an end-to-end process, focused from supplier relationship management to customer satisfaction, with products and services delivered through people, systems, and business partners. It is at this point where the constituents see a need for something that enables them to enter each other's databases and extract the vital information that assures positive results. Enter BPM! BPM becomes the technical breakthrough. It takes advantage of the ability to *componentize* within the software. By that we mean it provides the linked firms the chance to reach into the databases and extract that portion of the information that will help with the issue or problem being addressed. Visibility becomes feasible, as previously hidden information is now available to those who need it. Control across the people, systems, and organizations reaches a new high level, closing the gap between management intentions and execution and success. By means of directly executable business process models, the connected businesses leverage the best of their IT infrastructure.

BPM then produces dramatic improvements to the targeted processes, ending in better business results through lower costs, greater speeds, shorter cycle times, higher quality, and better service. Successful implementations typically begin as tactical solutions to specific problems, before being adopted as part of the processing. Using a strong process-based methodology and paying careful attention to the architecture become as important in this environment as selecting the right technology. An early pilot to create a proof of concept becomes a useful way to qualify the BPM tools being applied to specific problems and opportunities.

BUSINESS PROCESS MANAGEMENT AS THE BREAKTHROUGH

The challenges of integrating processes and knowledge within and across enterprises are best answered by a combination of:

- A visionary technical architecture:
    - To which the (extended) enterprise can transition over time
    - Which will allow the business to adopt new processes, applications, and technologies quickly and efficiently
    - Which avoid redeveloping large portions of legacy software
    - Without disrupting existing users
- Business process management systems provide:
    - A new generation of tools that put the management of business processes back with the business community, freeing the IT community to do what it does best — enable the processes with the best available technology

A VISIONARY TECHNICAL ARCHITECTURE FORMS THE FOUNDATION

As supply chain efforts progress, stovepipe thinking and point-to-point technical integration give way to flexible, business-process-based architectures. The complexity and diversity of enterprise systems, the growth of middleware, and the drive for the next level of efficiency and productivity, both within and across organizational boundaries, mandate process thinking. But unlike reengineering, today's processes must be directly executable and evolve incrementally, with minimal impact on business operations. It is for this environment that CSC e4^{SM} has been designed.

> The future enterprise will be complex, federated, connected, collaborative, dynamic, constantly evolving, and unpredictable.

CSC e4^{SM} provides the flexibility of approach that enables a firm to adapt its infrastructure to new business conditions (for example, acquisitions) and new technologies, without the need for wholesale changes and integration retesting. It has been especially designed to meet the future architecture challenges facing most businesses attempting to excel at extended enterprise processing. As the evolution of IT infrastructure and architecture continues, exploiting existing and emerging technologies, CSC e4^{SM} seeks to address many of the problems facing businesses and their supply chains today and bring answers to many questions:

- How does a business operate in a state of perpetual change and adaptation?
- How does an individual firm become an enabler of business process change while not throwing away the investments made in legacy systems?
- How does a company integrate IT following mergers or acquisition?
- How does a business operate with some of the world's largest companies while also gaining value from some of the smallest and more innovative firms?
- How can a firm provide a cost-effective, common multichannel Web access for users, suppliers, and customers?
- How does a business take cost-effective advantage of new access technologies?
- How does a business organization and its allies implement IT architecture that will dynamically grow with business and IT objectives?
- How are third-party applications brought in, without major business risk and upheaval?
- How can a business achieve "true" collaboration with business partners and service providers?

ANSWERS COME FROM THE NEW BREED OF IT ARCHITECTURE: CSC e4^{SM}

CSC e4^{SM} is a proven world-class, award-winning example of the new-style IT architecture used for ASCM — a breakthrough capable of integrating individual applications, whether new or within legacy systems — that enables the open flow of processes and all the attendant data between systems, across organizations, between enterprises, and among trading partners. CSC e4^{SM} can rapidly integrate Web-based applications, front- and back-office systems, enterprise resource planning systems, and package software applications. It verifies what can be accomplished using BPM as the link between extended enterprise business partners.

The architecture uses application adapters to allow true plug-and-play capability, banishing repetitive and costly point-to-point solutions. CSC e4^{SM} has been designed to be flexible, scalable, and rapidly deployed, so as to deliver a future-proof solution to extended enterprises for a low cost of ownership. Its flexibility allows business processes to be tuned or amended without any coding changes.

HOW CSC e4^{SM} WORKS

CSC e4^{SM} uses a component approach that integrates Web technology and legacy and commercial applications with a process management engine connected by adapters to middleware services, to deliver a business-process-oriented solution. The business logic of the adapters allows drag-and-drop capability when reengineering processes, which, along with CSC e4^{SM}'s true plug-and-play and vendor-independent flexibility, delivers a future-proof solution to any IT architecture. The key aspects of CSC e4^{SM} are shown in Figure 2.5.

These aspects come in layers and begin with the distribution layer, designed to provide multichannel access within the network. Users can interact through various channels, including browsers, PDAs, and mobile phones, enabling employees to use the channel best suited to their work and consumers to select the device of greatest convenience. CSC e4^{SM} allows the process designer to take advantage of the features of various channels. This layer also manages the physical connections, ensuring authentication, security, and eligibility of the source. The architecture creates an enterprise format, so important for cross-business communication, allowing users to interact across the full network and manage tasks associated with their process steps.

Moving through the enterprise portal to the business process engine, the process manager allows the business architects to describe the end-to-end processing using available services. It permits process design logic to be based on data captured from enterprise applications, while allowing rapid deployment of new and changed business processes. Design patterns can be reused as subprocesses within larger processes. The process manager can also establish the definition of synchronous, asynchronous, and parallel steps within a process.

The coordination layer manages the Web services and applications integration requirements. It marshals requests for services to the correct component interface, ensuring that responses occur in context with the overall process design. This layer provides services for current industry data and application format conversion between process constituents. Middleware technology is used to deliver all service requests and to connect applications.

Distribution

The distribution layer enables CSC e4[SM] users to interact with the enterprise through different channels such as browsers, PDAs, and mobile phones. This enables employees to utilize the channel most suited to their work or consumers the device of greatest convenience. CSC e4[SM] enables the process designer to take advantage of the different features and benefits of any channel. This layer manages the physical connection between managers; ensures authentication, security, and eligibility of the channel source; and clarifies user role against enterprise entitlements and access control lists.

CSC e4[SM] transforms requests from the channels to a channel-neutral, enterprise format for processing by the process interaction manager, through the portal interface. Users interact with the end-to-end process and manage tasks associated with the process, optionally supported by work flow management technology. Responses are translated back to the specific channel format.

Coordination

The coordination layer manages the Web services and application integration requirements of the enterprise. It marshals requests for business services to the correct application component interface and ensures responses occur in context of the overall process design. This layer provides services for on-the-fly industry data and application format conversion between process participants, together with process and systems monitoring and management of both the enterprise applications and CSC e4[SM] elements. Middleware technology is used to deliver all service requests and to connect applications.

Business Process Management

The process manager allows business architects to describe an end-to-end business process, or an interface to a business process of a partner, using services available in the business service directory and/or reusable process patterns in the process repository. It allows process decision logic to be based on process data captured during the process or returned from interactions with enterprise applications. The process design tool allows rapid deployment of new and changed business processes once all participants have been registered with CSC e4[SM]. Process design patterns can be reused as subprocesses within larger processes. The process manager allows the definition of synchronous, asynchronous, and parallel steps within a process, including the business-level definition of transactions or compensating activities.

Service

The service layer consists of the business interfaces of the enterprise applications known to the CSC e4[SM] environment. Each application within the service layer exposes itself to CSC e4[SM] using connectors, adapters, and process projectors, as required. CSC e4[SM] defines standards for such connectivity, enabling rapid and simple integration with other applications within the enterprise.

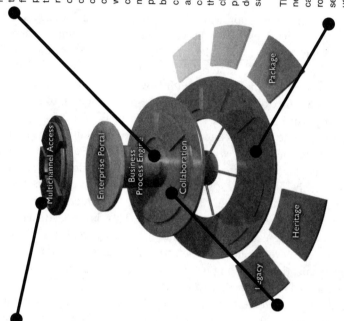

Figure 2.5. CSC e4[SM] Architecture

The service layer is occupied by the business interfaces of the enterprise applications known to the CSC e4^{SM} environment. Each application within the service layer exposes itself to the user connectors, adapters, and process projectors. CSC e4^{SM} defines standards for such connectivity, enabling rapid and simple integration with other applications.

For the value-managed enterprise, the significant aspect of this architectural approach is the separation of business process from application. Once this separation is achieved, processes can flow freely across organizational boundaries, within and across the enterprise. Data are accessed from whichever applications store the information and delivered after processing to those applications that need it — all without the users being aware of the applications at all. The users are free to do what they do best — manage and run the processes.

THE BENEFITS OF THE NEW ARCHITECTURES

New architectures, of the type exemplified by CSC e4^{SM}, provide reduced total cost of process and infrastructure ownership while increasing the strategic return on investment from deployment of reusable and manageable enterprise architecture. This translates into the following business and technical advantages:

Business	Technical
End-to-end process visibility, control, and accountability	Lower integration and operational cost
Rapid introduction of new business units and new products and services that are operationally dependent on automated processes	Resources freed up from legacy maintenance to focus on process improvement
Ability to manage process value end to end and extend those disciplines across all processes within and across the enterprise	Flexibility to add and remove application components to decrease time to market and exploit windows of opportunity
Focus on customer through strategy-driven business process design	Technical barriers to business collaboration and change eradicated
A renewed enterprise architecture under the control of the business	Facilitate restructuring and repurposing of legacy applications

BUSINESS PROCESS MANAGEMENT SYSTEMS BECOME THE KNOWLEDGE LINK

At the heart of the new architectures is BPM. BPM is an emerging technology that heralds a replacement for the painful experience of past reengineering projects, where processes were redesigned in a one-off exercise, which led to extensive and costly systems and organizational replacement and disruption. Such programs of work are no longer acceptable to management teams, except in the most extreme cases. The focus is now on continuous process improvement enabled by BPM and business process management systems (BPMS).

> "Business Process Management is THE thing...not integration, not messaging. It goes far beyond integration and hooking up applications.
> It is the *essence* of good business."
>
> Jim Sinur, Gartner Group
> July 2002

The business processes that are crucial to an enterprise's strategy are by their nature both complex and dynamic. "Complex and dynamic" means the processes have many steps and participants, and the participants are free to run a process in a way that is best suited to the situation they are in. This condition is illustrated in Figure 2.6. Until now, available technologies have been unable to deal with these processes as they would ideally be executed. Instead, they have dealt with them by dumbing them down, making them less strategic by either remodeling the processes to fit into a package or leaving them as manual processes. Package processes are less likely to allow a company to distinguish itself from its competitors. Rules-based processes are limited to the number of different situations that the process designers can dream up and cater to; anything else is an exception, to be handled outside the process.

> "Enterprises should begin to take advantage of explicitly defined processes. By 2005, at least 90% of large enterprises will have BPM in their enterprise nervous system (0.9 probability).
> Enterprises that continue to hard-code all flow control, or insist on manual process steps and do not incorporate BPM's benefits, will lose out to competitors that adopt BPM."
>
> Gartner, *Wall Street Journal Business Process Management Review*, June 2002

BPM for the first time lets businesses take full ownership of their processes without the limitations of underlying technologies. BPM does not so much change processes — which are the things people do — as allow people to change them by providing a supporting technology that is more flexible than anything that exists today.

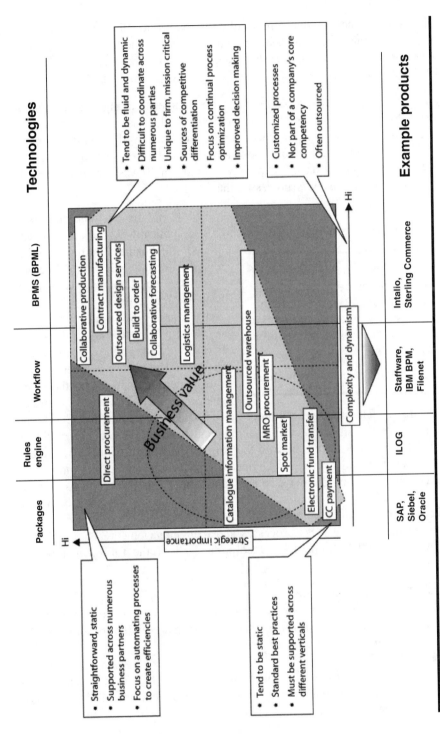

Figure 2.6. Complex and Dynamic Processes

BPM is a framework that consists of tools and services that provide real-time visibility, continuous management, and optimization of end-to-end business processes that interact with people, systems, and across organizational boundaries. The BPM methodology is based on the following set of assumptions:

- Business processes operate in a state of perpetual change and adaptation.
- Processes interact with other processes and cross-cut each other.
- Processes can be distributed, end to end, long-lived, collaborative, and transactional and involve several participants (people and systems).
- BPM is based on a technology solution, a BPMS, to create directly executable business processes that evolve.

A BPMS enables companies to model, deploy, and manage mission-critical business process, that span multiple enterprise applications, corporate departments, and business partners — behind the firewall and over the Internet. It provides a cycle of continuous feedback and improvement for the end-to-end business process life cycle.

> "Business processes have been around since the beginning of business. Business Process Management Systems are the next step in making them explicit, executable and adaptable."
>
> **CSC Research Services**

Central to a BPMS is Business Process Management Languages (BPML). There are a number of standards emerging, with which these languages comply. These are illustrated in Figure 2.7. The supply chain processes defined, for example, in SCOR® and collaborative planning, forecasting, and replenishment (CPFR) are all definable in the BPML. This feature goes to the core of the benefit of these new tools. BPMS is not business process reengineering, enterprise application integration (EAI), work flow, or another package application. It is the synthesis and extension of all these technologies into a unique whole. This unified whole becomes a new foundation upon which the enterprise is built, one more in tune with the nature of business processes and their management.

A complete BPMS platform incorporates all of the following types of functionality:

- EAI and middleware
  - □ BPM complements EAI tools as a means to create new business processes and services that the EAI and middleware tools are designed specifically to integrate.
- Work flow
  - □ BPM encompasses work flow. However, while work flow addresses document routing and facilitates manual work, the BPM engine also encompasses pure automation and automated decision making.

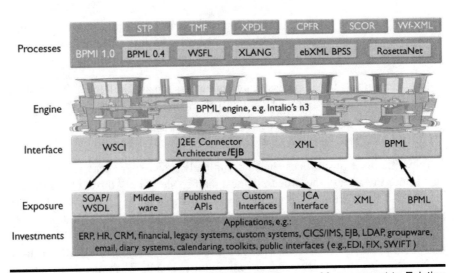

Figure 2.7. Emerging Standards for BPML Bring Process Management to Existing and Future IT Investments (Source: Intalio and CSC)

■ BPM tool set
   □ A unified set of process tools from visual process modeling to process execution.
■ Business intelligence and process management functionality
   □ Support for business activity monitoring that enables process performance reporting and analytics.

BUSINESS PROCESS MANAGEMENT OPERATES WITHIN THE NEW ARCHITECTURE

Attention can now be turned to the advantages for the supply chain provided by the combination of a new-style architecture with a BPMS at its heart. Change is a fundamental driving force in business, and therefore agility is a mandatory requirement of enterprise architecture. A BMPS streamlines internal and external business processes, eliminates redundancies, and increases automation, providing end-to-end process visibility, control, and accountability. This improvement is not limited to the processes in a department or a business unit. Processes that comply with the new standards can join with other similarly compliant processes wherever they are — in other parts of a single enterprise or with value chain partners across the whole supply network.

The new BMPS, embedded at the heart of a new architecture, support the automation and continuous improvement of traditional processes such as marketing and sales, human resources, finance, operations, supply chain, product design, forecasting, logistics, and customer relationship management, as well as:

- Commonly used process design patterns such as CPFR, SCOR®, straight through processing (STP), and Telemanagement Forum (TMF)
- Industry standards such as XML, J2EE, and .net
- Open standards like GPL, JCA, and BPML

Now the user of a process has a completely new experience. The computer screen used to access the process is designed to show only those elements that are needed to execute the process — even if the data being processed are in several different applications. The steps in the process are presented precisely the way the user finds them in the easiest manner. If she or he wants to change the process for the better, the business user can do so, in the BPMS, usually without having to involve the IT community at all. The major advantages include:

- Help for the business user, who knows, owns, and can upgrade the process
- Minimize the time in executing the process — no more logging on and off different applications until the job is done
- Extend the process across departmental and company boundaries, so the environment for collaboration is set up, ready to go

Other questions can be adequately answered as well, like what's in it for IT. To start, the IT community can implement a new-style architecture and provide functionality to the user more quickly than before. Applications are implemented in their vanilla form, and processes and screens are developed more quickly in the BPMS. Fewer user screens are needed because processes are executed in fewer steps. Once the new architecture is up and running, application upgrades can be made at the time that suits the IT community and are totally transparent to the user community, so maintenance costs are reduced.

"...the average F500 firm spends over $200 million annually trying to digitize their business and improve productivity by patching processes untouched by their current enterprise applications."

eAI Journal

THE VALUE OF TRUE BUSINESS PROCESS MANAGEMENT

For the IT community, typical benefits of the new-style architecture and BPM include:

- Reduction in IT development cost by up to 30%
- Payback typically between six and nine months
- Typical enterprise integration project implementation down from 18 months to 7 months
- BPM supports highly iterative IT projects; start small then grow

For the business community:

- Some early adopters have achieved annual revenue increases of 5%, with a three-year return on investment as high as 250%.
   - These figures coexist with annual IT cost reductions of 10% or more and savings of 50 to 80% for application management alone.
- BPM puts businesspeople back in the driving seat and enables and encourages continuous business process improvement and evolution. Directly executable business models become feasible, with reusable building blocks for the business, so as the process library builds up, subsequent processes become easier and quicker to define and implement.
- BPM aligns processes more directly with business objectives and streamlines internal and external business processes, eliminating redundancies and increasing automation.
- BPM provides end-to-end process visibility, control, and accountability, supported by common process integration protocol (BPML) behind the firewall and over the Internet — ideal for supply chain collaboration. BPM allows for global process convergence while catering to essential local customization.

SUMMARY

As firms progress along the maturity model and approach ASCM characteristics and the advantages of sharing knowledge across an extended enterprise, they overcome two limitations: the need to share vital information within and outside of the organization and the methodology by which the sharing can be done easily and economically. BPM becomes the enabler to help firms extend their efforts into the advanced levels and selectively share knowledge of pertinence to the processing taking place. With a BPMS selected by the constituents involved and designed to create satisfaction for the key customers and to optimize the process steps from key suppliers to delighted consumers, the advantages can lead to market dominance. The necessity is to form a working group of business allies dedicated to finding the correct processing, enabled by an active BPMS.

3

THE PROCESS-BASED PROFIT AND LOSS STATEMENT AND BALANCE SHEET

At this point, we have introduced several ideas:

■ A supply chain maturity model can be used to calibrate a firm's position with regard to advancing to the highest appropriate level and to determine the order-of-magnitude benefits for making such a progression.

■ The SCOR® model can be applied to match the best practices across an advanced progression with the elements of that model.

■ A series of matrices can be used to determine that the best practices have been achieved at the desired level of maturity.

■ Customer satisfaction, delivered through an intelligent value network, will be a major factor distinguishing the businesses linked in a well-planned and -executed extended enterprise.

We have also indicated that business process management and business process management systems should be used to make the communication breakthrough between important constituents of the supply chain network and to gain access into disparate technology systems, so that business allies can access the necessary knowledge across the enterprise, of which they are one link, to make important end-to-end process changes. With those concepts as our background, in this chapter, we will introduce the central idea behind this text: Results from successful supply chain and networked processing efforts should be documented and traced directly to the financial performance of the firm at the center of the network and its collaborating business allies. In this sense, advanced supply chain management (ASCM) and business process management become enabling factors in achieving higher financial performance.

Our basic premise is that once the appropriate connections are made between business processes within the network, so a view of potential optimized conditions and highest possible customer satisfaction can be attained, attention should turn to pursuing the necessary improvement effort while verifying the enhanced financial performance. This means that the real opportunities for increasing value move from nebulous concepts to identified additions to revenues and profits. This thesis is encapsulated in the transition values for each process in the maturity model introduced in the first chapter. By identifying a firm's and its enterprise's current and desired status on the maturity model and the SCOR® matrices, and relating the examples, improvement percentages, and actual numbers to be documented through case studies to one's own financial statements, the reader will be able to create a first-cut value-based transformation plan.

That means the processes being improved are cared for not simply because they are elements of the supply chain, but because the improvements can be tracked to financial performance. This idea is no longer part of an elusive quest. Recent reports by major companies are beginning to verify exactly what can be accomplished. In one specific example, IBM Corporation determined that its 2000 financial statement showed an inventory of roughly $4.8 billion, while sales were $88.4 billion. Using a formula for days of inventory, which divided the amount of inventory by the sales and multiplied the result by 365 days, the company reckoned its days of inventory at 19.7 days. A one-day reduction to that figure ($4.8 billion divided by 19.7 days) would generate a one-time positive cash flow of $241 million.

For further substantiation of the potential for ASCM, a few examples verify what can be brought to the balance sheet and profit and loss (P&L) statements:

- Gillette Company, the Boston-based manufacturer of shaving and oral care products and Duracell batteries, has reduced its inventory by 30%, or $400 million, since the January 2000 formation of its supply chain organization.
- Quaker Oats, a division of Pepsico, Inc., through its supply chain initiative dubbed North American Manufacturing Study, expects to achieve savings between $60 and $70 million.

- KYB Manufacturing, a Franklin, New Jersey supplier of automotive struts and shock absorbers, experienced a volume explosion from 80,000 units per month to over 500,000. Applying supply chain techniques and enabling technology to track inventory, order parts, and forecast demand, the company reduced raw material inventory from $7.8 million to $6.3 million. Scrap has been reduced by 50%, while errors have been eliminated and on-time rates have skyrocketed.
- Unilever, the global consumer goods firm, through its Path to Growth supply chain effort, expects an annual sales growth of 5 to 6%, coupled with a 16% increase in operating margins.
- IBM's Integrated Supply Chain Group is expecting to cut up to $2 billion from its $40 billion raw material and supplies spend.
- Johnson & Johnson Medical, Inc. saw an increase in forecast accuracy from 12% to 53%, a reduction in inventory turnaround from 154 days to 110 days, and an increase in customer service from 95% to 99.38%.

These types of documentation serve to verify that the search for real financial gains is not a spurious effort, but is bringing the long-sought improvements to the P&L statement and balance sheet.

THE EFFORT BEGINS WITH DRIVING METRICS

As we have worked with numerous firms in various industries and helped guide them toward enhanced performance through a concerted supply chain effort, a variety of methods have been used to track the actual improvements, as well as a number of specific measurements. Depending on the orientation of the company, we have been asked to document enhancement to economic value added, earnings per share, return on net assets employed, return on investment, improved cash flow, and better earnings before taxes. In virtually all cases, we were able to show the direct link between supply chain improvement and the desired metric.

Unfortunately, along the way we encountered a number of firms that had difficulty making the direct connection between important financial measures and what was resulting from their supply chain actions — in a way that could be reported on their earnings statements. In spite of good efforts to improve sourcing costs, cut transportation and logistics costs, reduce dependency on inventory, implement better forecasting and planning and scheduling, introduce greatly improved order management systems, and enhance sales lifts through better manufacturing and delivery techniques, many companies simply lack the ability to show the direct association with improved financial performance. It is our intention to eliminate that cloud from the supply chain environment and show how such a linkage can be established and tracked.

As another example of how the potential positive benefits can be derived if a driving metric is at the center of the effort, we would cite the actions taken by Colgate-Palmolive (C-P). This firm has concentrated on achieving the highest market share in one very competitive market, oral care, which includes toothpaste and associated products. From a position in 1994 where the firm had a share of 21.9%, C-P has progressed to a point where it now has 32% of that market. As this performance was increasing, C-P was deep into a supply chain effort, with one concentration on *cash-flow return on investment*. That metric stood at 14.4% in 2003, versus an industry average 4.8%.

While part of the transformation was no doubt due to significant product innovations, it was dramatically impacted by the connection to C-P's supply chain effort. According to one study, "This let Colgate integrate demand and supply chain information — and decisions — more effectively. After years of stagnant sales and narrowing margins in the mid 1990s, Colgate rethought the entire process involved in its value chain. The company then slashed supply chain costs for all its products even as it shortened manufacturing and delivery times." The report further states that "Colgate's planning is now much more precise: Errors in forecasting have been cut from 61% to 21%, and case-fill rates nudged up from 94% to 97%" (Koudal and Lavien, 2003, p. 82).

To make the most of the exposition to be delivered, it is best for the reader to set out those metrics of greatest importance to his or her business. Table 3.1 is an excellent reference in this area. Created by Raymond Clayton, and reported in *Supply Chain Management Review,* it presents the kind of financial measures of most importance to a business. Next to each measure is a brief list of the elements from the financial statement that are related to the measure. Then Clayton lists the performance drivers that need improving in order to enhance the desired performance (Clayton, 2002, p. 35). Each company is advised to develop its own set of financial outcome measures and performance drivers. With the Clayton chart as a guide, the information we present will then have the most benefit in determining how a concerted effort will produce the type of desired impact.

Along the way, it is equally important to understand the relationship between financial results and the specific elements within the supply chain that impact performance in both directions — positive and negative. This refers to the need to make sure efforts devoted to supply chain improvements are linked with process changes that are valuable to the firm, and not just measures that show improvement without really having a positive effect on financial performance. For example, we could consider the very important need to segregate suppliers and customers by value, before developing supply chain changes that impact both areas, so a lot of effort is not wasted on areas that have little additional merit (weak suppliers and customers serviced at a loss).

We note, for example, that one firm which has made great progress with supply chain improvements, Procter & Gamble, is constantly faced with the need to put emphasis in its supply chain effort on areas that have the most overall impact to the firm. According to one report, "In 2002, just 12 of P&G's 250-odd brands generated half of its sales and an even bigger share of net profits" (Andrew and Sirkin, 2003, p. 77). Clearly, a lot of valuable effort could be wasted on brands that have little overall merit. It becomes essential, as a supply chain effort matures, to track what return is being made from the specific areas of most value, so that carefully directed investments in time and resources are made in the areas of greatest importance.

SUPPLY CHAIN IS ABOUT MORE THAN REDUCING COSTS

To begin, then, the process of tracing value to the P&L and financial statements, we must accept the premise that there is more to a supply chain effort than simply cutting costs in perpetuity. That sort of intention will get the effort launched and provide the early savings to fund a continued effort, but it cannot be the long-term element of sustenance. At its core, ASCM is concerned with improving business processes, and thereby enhancing a variety of important metrics, including those affecting costs, revenues, asset utilization, customer satisfaction, and return to stakeholders. In Figure 3.1, we see the business process challenges. They are to improve all of the important enterprise processes.

Using SCOR® as a guide, we see the linked supply chain processes, laid over the overall enterprise processes with a few carefully selected additions. Beginning with the source or buy process step, the challenge begins with a need to move beyond just finding the lowest priced suppliers, to work with key suppliers to help with the design and development of new products and services. The intention becomes to create and deliver a flow of innovative products and services in the shortest possible cycle time, from concept to commercial success, while improving the percentage of successful efforts. The next need is to link the development processes in a way that features of customization can be added for the highest priority customers and consumers.

Underpinning the third important process step is the need to have a communication system at work that reliably informs the constituent members of the supply chain network of exactly what is happening within the overall system.

Table 3.1. Linking Financial Outcome Measures and Purchasing and Supply Chain Management Performance Drivers

Financial Measures	Elements of Financial Statement	Purchasing and Supply Chain Management Performance Drivers
Growth	Revenue Sales Market share	■ Supplier's quality, delivery, flexibility, and innovation ■ Supply management's involvement in product development ■ Outsourcing decisions ■ Price and quality of purchased goods
Profitability ■ Earnings per share and economic value added ■ Stock value ■ Profit margin	Revenue	■ Quality improvement and cost reduction of finished goods ■ Inventory management to support desired customer service levels ■ Order cycle reduction ■ Quality management
■ Cost reduction/savings	Cost of goods sold ■ Purchase ■ Freight in ■ Direct labor ■ Indirect expense ■ Inventory cost ■ Depreciation on equipment and plant	■ Cost savings and reductions in purchased goods ■ Improved supplier selection and management ■ Supplier development ■ Total cost of ownership analysis ■ Reductions in supply management's operating expenses ■ Improved inventory management ■ Warehouse cost savings ■ Transportation cost savings ■ Outsourcing to reduce capital investment and price ■ Lease or buy decisions
	Other operating costs ■ Amortization ■ Contract labor ■ Depreciation ■ Insurance ■ Interest ■ Maintenance ■ Office expense ■ Travel ■ Utility	■ Cost savings/reduction from supply management operations ■ Standardization of purchasing function's structure and processes ■ Electronic data interchange and other cost-effective ways of doing business with suppliers ■ Vendor-managed inventory ■ Supply management's operating expense ■ Supply management's administration expense ■ Outsourcing rather than purchasing capital equipment ■ Financing costs, including inventory financing, facility financing, and technology financing

Asset management
- Return on asset
- Return on investment
- Return on net asset
- Return on capital
- Return on capital employed
- Return on equity

Inventory ■ Raw material ■ Work in process ■ Finished goods	Major decisions on outsourcing issues and inventory management of raw materials ■ Raw material management ■ Order cycle management ■ Inventory management ■ Vendor-managed inventory ■ Contract manufacturing
Accounts receivable	Quality and price differentiation to increase the product's competitiveness, so that better payment terms can be achieved from customers
Fixed assets	Support outsourcing decisions and capacity needs ■ Property ■ Plant ■ Warehouses ■ Storage and handling equipment ■ Purchasing systems hardware and software ■ Communications equipment
Accounts payable	Payment terms with suppliers ■ Payment/discount terms with suppliers ■ Cost savings/reduction from suppliers

Cash management
- Cash flow
- Cash cycle

Operating activities	Cost savings ■ Inventory management ■ Order policy ■ Payment terms
Investing activities	Make or buy decisions ■ Outsourcing decisions ■ Lease or buy decisions
Financial activities	Payment terms

Source: Adapted from R. A. Clayton, *Purchasing: Pipeline to Profits*, Clayton Advisory Management Practice, reported in *Supply Chain Management Review*, November/December 2002, p. 35.

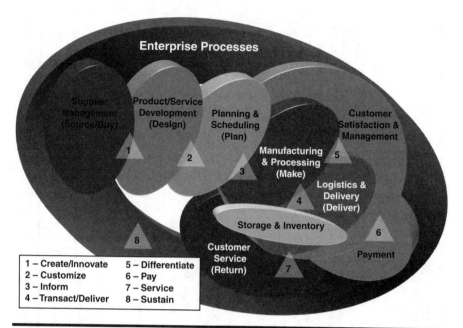

Figure 3.1. The Business Process Challenges

That means there is visibility into the linked partners' operations and access to their databases in a way that vital knowledge can be easily extracted, without risk to security. This information becomes a key element in effective forecasting and planning, so that schedules going to the manufacturing process are truly linked with actual demand.

Once into the manufacturing or processing area, the business allies work together to bring the transact and deliver processing to the highest possible level of effectiveness. In Chapter 7, we will describe how some advanced firms are combining lean manufacturing techniques with ASCM, Six Sigma quality, and the SCOR® model to reach the best possible level of response. The result leads to the fifth link in the challenge, which is to differentiate the network response from that of competing systems. Here we see the need to have storage and inventory under network visibility and control, so the right goods are at the point of need at the right time, in the right quantities and condition. With a network focus on this critical area, the inventories are not moved around in the supply chain, but are eliminated until only the absolute correct amount of goods is reached, with very little buffer stocks to cover unexpected conditions.

With the best possible level of customer satisfaction and customer management being achieved through the network's differentiated processes, the linkage moves to the payment process, where shorter cycle times and extremely accurate levels are marks of the actions. If there are needs for returns, they are handled quickly and effectively, as the customer service processing is also at the highest levels. Finally, we see the need to sustain the processing, through constant and mutual attention by the involved network partners to always keep the individual process steps at leading conditions.

The kind of values to be added along these challenges has been presented in the maturity matrices. What a firm needs to do is decide how far along the continuum it is important to progress, what business allies will be needed in its network to make the journey, and then to establish the benefits that will be delivered and a means of capturing the actual value added to performance statements. That is the essence of what we are now considering. For example, a manufacturing firm could determine that one important link is between improvements to supply chain operations affecting delivery of finished goods and financial performance in terms of cash flow. The necessity becomes to clearly show the connection between actions taken and elements on the balance sheet, explaining working capital tied to inventory levels and the impact on carrying costs. Supply chain efforts aimed at inventory reduction, including vendor-managed inventory, just-in-time purchasing, and continuous replenishment programs, are all examples of techniques that should show a direct improvement. Links could also be made to improved revenue through sales garnered as a direct result of improved supply chain distinctions or to improvements made in transportation costs for final delivery of the goods to customers.

From another aspect, a firm could show the direct link between a more effective sales and operations planning process and better inventory management or a decrease in lost sales. The first link impacts the need for working capital and enhances the return on assets employed, while the latter improves the top line revenues. Improved planning has many other benefits, including reduction of missed sales opportunities due to better product availability and higher fill rates. These elements are harder to track but do result in increased revenues and market share. Reliable studies have shown estimates indicating that as much as 4 to 5% of revenue could be lost to not having the right product available at the time of need. Our work with several consumer products firms has indicated that sales lifts as high as 5 to 7%, due to improved matching of actual demand with capability to deliver within the responding system, are possible.

Other process steps directly reduce elements of manufacturing and delivery costs, lowering the reported cost of goods sold and increasing operating margins. A positive hit to operating margin, moreover, has a direct and positive impact on earnings per share and shareholder return. It is not unusual in our world of interaction with companies and their supply chain efforts to see benefits of eight to ten times or more the actual investment in supply chain efforts.

The most common improvements are new revenues gained through distinctions created in the supply chain, lower transportation and storage costs, less working capital and lower carrying cost tied up in excess inventories, improved operating costs gained through better planning and manufacturing, and less sales and administration costs due to improved efficiencies and elimination of manual processing. As a supply chain effort matures, other factors receive attention and the list increases. Generally, we find attention goes to increasing earnings per share and asset management, and thereby enhancing stakeholder value.

Our general findings have been that while the major emphasis in the early levels of the maturity model is placed on cost reductions in sourcing and logistics and what often can be nebulous inventory reductions, the equal or greater benefits can be derived in later levels from improvements to revenue and sales margins (by virtue of showing the values added to customers through the distinctions gained in the supply chain efforts). We also see an increase in the return on investment, as the effort is sustained and extended through collaboration on new product and service introductions and better sales event efforts. As the collaborating businesses also look at the assets employed and decide which business ally has the best capability, asset transfer or asset ownership becomes an element for reducing costs tied up in such assets and balance sheet reporting.

THE APPROACH BEGINS WITH MAKING THE FINANCIAL LINK TO PROCESSES

The value created in a supply chain network can be measured in several ways, but the key is to find a set of measures that encourage the efficient creation of profit. These measures look good if either profit goes up or the assets used to generate the same amount of profit go down. We shall begin by using the concept of return on net assets (RONA) as it relates to supply chain management. This technique will take us well on the journey, through levels 1 and 2, and will support a more advanced treatment when we get beyond level 3.

RONA is the ratio of the profit earned to the assets used to produce it, as shown in Figure 3.2. As illustrated, it uses factors related to supply chain to arrive at a reasonable measure of return. In other words, it compares the P&L account with elements of the balance sheet. The value of RONA is that it allows a balance to be struck between the often-conflicting pressures of trying to increase profit and manage the asset base. For example, if you increase your inventories, you may be able to sell more, but will you make commensurately more profit? If RONA goes up, you will. If it goes down, you won't.

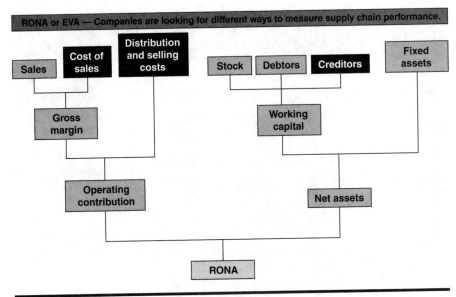

Figure 3.2. Supply Chain Performance Measures

A second consideration is to choose the scope for applying RONA. Few supply chains involve the whole of a multidivisional corporation working as a single entity to supply its customers. In fact, from the authors' experience, most corporations can be divided into autonomous horizontal businesses, each with its own self-contained group of products and set of assets that produce those products, as shown in Figure 3.3. Each operates with its own supply chain and goes to its specific markets. The horizontals may share suppliers and channels, but the key decisions about range of investment management, investment in assets, and tactical decisions around sales order management can all be taken within the horizontal unit, with no great conflict with other horizontals. The key determinant of a good decision is whether RONA for the horizontal organization will go up or down. Interestingly, these horizontals often cross the formal boundaries in the organization and are the root cause of some of the complexities encountered in supply chain management.

These horizontals will be the basic building blocks for our thesis. A self-contained level 1 organization will be able to decide on its product portfolio and investment in assets. It will have a supplier base, which it may share with other parts of the corporation to which it belongs and will be able to see the sales it generates from its (possibly shared) channels to market, and will incur a share of the cost of those channels. But for our purposes, it is a self-contained business, governed by RONA.

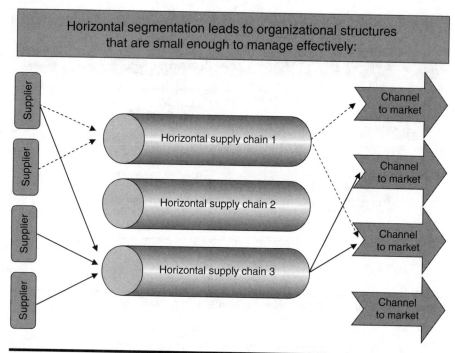

Figure 3.3.   Horizontal Supply Chains

The process model introduced in Chapter 1 is the process model for our self-contained horizontal organization, shown here again as Figure 3.4. Now we have an enterprise with processes and a way of measuring if those processes are improving. The question that needs answering is: As processes improve, and we move up the levels, how much value is created? To answer this query, we need to think about a process-based set of measures — a process P&L account and a process balance sheet.

THE PROCESS-BASED PROFIT AND LOSS ACCOUNT AND BALANCE SHEET

Figure 3.5 shows a typical P&L account. Income of £1 million generates a profit of £100,000. The statement shows how our business is doing overall, but does not tell us how well the processes are performing. What are the costs associated with the Plan process? With the Make or Deliver process? Are these funds being spent well?

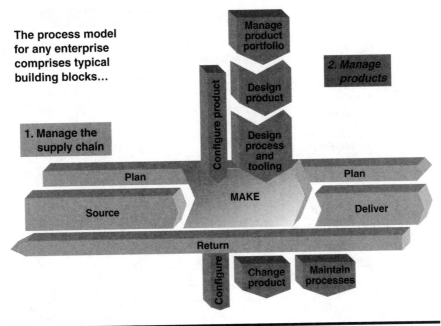

**The process model
for any enterprise
comprises typical
building blocks...**

Manage
product
portfolio

*2. Manage
products*

Design
product

**1. Manage the
supply chain**

Design
process
and
tooling

Configure product

Plan

Plan

MAKE

Source

Deliver

Return

Configure

Change
product

Maintain
processes

Figure 3.4. Process Maturity: Managing Supply Chain (Based on Supply-Chain Council SCOR® Model)

Taking each line of the P&L account, Figure 3.6 shows a possible allocation to each of the five supply chain processes. The product design processes can be dealt with in a similar way and can easily be added to the figure. The key to the decisions is deciding who is accountable for generating the income or consuming the cost — or has some responsibility for the process. As in many situations, these roles are not always clear. In the authors' experience, one process should be accountable for each key metric or share of it, and the contributions of others should be clearly understood. The accountabilities and responsibilities are then built into individuals' performance measurements.

The same treatment can be given to the balance sheet, as illustrated in Figure 3.7. Of course, accountability for some of the balance sheet items will lie with other process owners outside of the supply chain. For example, loans and long-term debt will be the province of the financial function. Nonetheless, they will all, ultimately, come under the control of one or more processes in the complete enterprise process map. The ultimate responsibility of the managers of the enterprise is, of course, to maximize the return for the shareholders and increase the strength of their balance sheet.

		£'000		
1	**Income**			
2	Turnover			1,000
3	**Cost of Sales**			
4	Materials	459		
5	Wages	213		
6	Total (4 + 5)		-672	
7	**GROSS PROFIT** (2 – 6)			**328**
8	**Distribution**			
9	Marketing	25		
10	Delivery Costs	26		
11	Total (9 + 10)		**-51**	
12	**Expenses**			
13	Rent/Lease	57		
14	Insurance	4		
15	Professional Fees	13		
16	Utilities	15		
17	Debt Write Off	13		
19	Motor	34		
20	Interest Charges	21		
21	Depreciation	21		
22	Total (13 to 21)		**-177**	
23	**NET PROFIT BEFORE TAX** (7 – 11 – 22)			**100**

Figure 3.5. The P&L Account

We now have a set of metrics against which the performance of each process can be judged as investments are made in increasing maturity. The same principles can and should be extended to nonfinancial key performance indicators such as fill rates, on-time delivery, and customer satisfaction. The impact of process changes can now be judged against the RONA performance of the whole enterprise. Are we reducing cost or increasing revenue? Are we investing in our balance sheet in a way that enhances our overall return on net assets? Are we seeing the kinds of returns promised by those who advised investing in supply chain efforts?

The task of establishing an internal supply chain and measuring its effectiveness is well illustrated by work done with Zeneca Agrochemicals, now part of Syngenta. The project piloted a new approach to supply chain management in the company's insecticide/fungicide suspended colloids business. Farmers around the world use these products to guard their crops from infestations by

Key: ✔ Accountable * Responsible		Plan	Source	Make	Deliver	Return
1	Income					
2	Turnover	✔	*	*	*	*
3	Cost of Sales					
4	Materials		✔	*		
5	Wages	✔	✔	✔	✔	✔
6	Total (4 + 5)					
7	GROSS PROFIT (2 – 6)					
8	Distribution					
9	Marketing	✔				
10	Delivery Costs				✔	✔
11	Total (9 + 10)					
12	Expenses					
13	Rent/Lease	✔	✔	✔	✔	✔
14	Insurance		✔	✔	✔	
15	Transport Costs	*			✔	✔
16	Utilities			✔		
17	Debt Write Off			✔	✔	
19	Interest Charges				✔	
20	Obsolescence	✔				✔
21	Depreciation	✔		*	*	
22	Total (13 to 21)					
23	NET PROFIT BEFORE TAX (7 – 11 – 22)					

Figure 3.6. The Process-Based P&L Account

insects or fungi. The product group was chosen because it formed one of more than 20 self-contained horizontal businesses that were identified in a previous study.

An early problem faced by the team members was getting their hands on any important supply chain data. The difficulty that the team faced was that most of the information in the business was closely guarded by the various functions involved in the processing. There was no supply chain information collected together to give an overview of what was happening to this whole piece of the business. Creating a map of the end-to-end supply chain and reaching to the root causes of the impediments eventually solved the problem. We set about mapping the supply chain, the result of which is shown in Figure 3.8. Information was collected on:

- The flow of materials into the process, including lead time and volume
- The inventory before the process

Key: ✔ Accountable * Responsible			Plan	Source	Make	Deliver	Return
		2003					
		£'000	£'000					
1	**Fixed Assets**							
2	Tangible Assets		200	✔		*	*	
3	Intangible Assets		50	*				
4	**Current Assets**							
5	Stock	150		✔	*	*	*	*
6	Debtors	320		*	*		✔	
7	Cash Bank/In Hand	90		*	*		✔	
8	**Total**	560						
9	**Current Liabilities**							
10	Creditors/Suppliers	410		✔			*	
11	Loans/Bank	60		*				
12	**Total**	470						
13	**Net Current Assets** (8 – 12)		90					
14	**Total Assets** (2 + 3 + 13)		340					
15	**Creditors**							
16	Amounts due after 1 year		25	*	*			
17	**Total Net Assets** (13 – 16)		315					
18	**Capital & Reserves**							
19	Profit & Loss Account		305					
20	Share Capital		10					
21	**Shareholders** (19 + 20)		315					

Figure 3.7. The Process-Based Balance Sheet

- The lead time through the process
- The utilization of the assets
- The value added during the process
- The outbound inventory, on its way to the next process in the chain

The next step was to create a RONA model for the value chain, as illustrated below:

Sales	100
Gross margin	70
Less direct costs	(20)

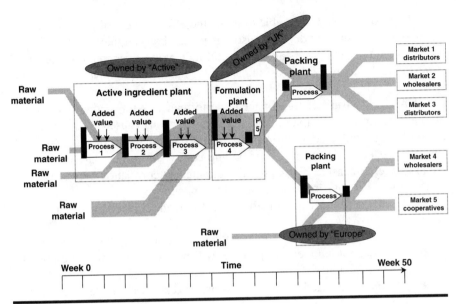

Figure 3.8. Supply Chain Map

Less indirect costs	(15)
RONA contribution	35
Fixed assets	30
Stocks	50
Debtors	35
Creditors	(25)
Assets employed	90
RONA %	39%

Now any changes to the processes of planning the end-to-end supply chain, including sourcing materials and scheduling production, could be judged absolutely for their impact on costs or inventory and relatively in terms of their impact on RONA. Local reporting became unnecessary and all decisions were taken across functional boundaries, for the good of the whole supply chain.

EXTENDING THE CONCEPTS BEYOND THE SINGLE ENTERPRISE

So far we have looked at a single enterprise, with its accounts and shareholders. This will stand us in good stead in level 1 and level 2. When we get to level

3 and beyond, we need to be able to judge the effectiveness of our investments not only on our shareholders but also our value chain partners. RONA is still a good measure, but there is another useful indicator — economic value added (EVA), which factors in the cost of capital for the different players in the value chain.

EVA is a measure of true financial performance, with particular emphasis on the efficient use of capital, that corrects distortions caused by anomalies in generally accepted accounting principles and takes account of the cost of all capital invested in an enterprise, including equity. More than any other traditional financial measure, EVA provides the highest correlation between corporate performance and increase in shareholder value.

EVA will increase if one or all of the following strategies are adopted:

- Operating profits are grown without additional capital investment.
- New capital is invested in any and all projects that earn more than the cost of capital.
- Capital is diverted from business assets that do not cover their cost of investment.

EVA is calculated by multiplying the economic book value of a firm's capital by the difference between the return on capital (r) and the cost of capital (c):

$$EVA = (r - c) \times \text{economic capital}$$

For example, if:

- Operating profit is £1,000 (usually net operating profit after tax)
- Economic capital is £3,000
- Cost of capital is 20%

then,

- $EVA = (£1,000/£3,000 - 20\%) \times £3,000$
- $EVA = (33.3\% - 20\%) \times £3,000$
- $EVA = £400$

EVA is a less intuitive measure than RONA, but becomes essential as we look at whole value chains, where the players will almost certainly have different costs of capital. Investment funding opportunities can take different shapes among companies with varying costs of funds. Now we can look at pan-enter-

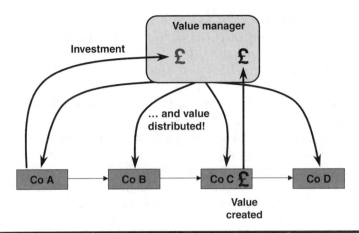

Figure 3.9. The Investment Process

prise processes and create what amounts to a value chain P&L account and balance sheet, by adding up the P&L and balance sheets of the constituent parts. The judgment on increasing overall performance can now be made using EVA as the overarching value chain metric. Managing transactions in this environment is reasonably straightforward: cash flows from the final customer back along the value chain. The major consideration is the profit taken by each company at each stage.

Funding decisions are more complex, as shown in Figure 3.9. To maximize value in the whole network, it is quite likely that an investment made by one company, say Company A, results in value in a second, Company C, farther down the network. Of course, the investments will be in improving the processes that cross the boundaries between the partners. An investment in equipment to improve the Make process in A may result in the reduction of material costs for the Source process of C. Until the thinking of level 4 and 5 comes into play, there are barriers to making this type of investment. There must be trust between the value chain partners that value will be shared, and there needs to be a mechanism in place to calculate the value and share it.

THE VALUE OF INCREASING PROCESS MATURITY

We now have the foundation for assessing the value of increasing process maturity. Based on an analysis of 40 case studies, we asked the owner of each case study to:

- Select the processes that had been improved
- Assess the maturity level that each process had been moved from and to
- Allocate the value created to the processes

The results are shown in Figures 3.10 through 3.13. Quite expectedly, we discovered that there were bona fide savings associated across all levels in the maturity model and for all of the leading indicators. Surprisingly, we could find little hard information about the returns process — perhaps indicative of just how recently this process has become important.

Doing a better job of planning showed the greatest return on inventory, as the need for extra safety stocks was reduced and redundant inventories were eliminated. Labor costs went down nicely, and customer service ratings increased. Sourcing showed a wide range of advantages, as this core function usually provides help throughout the maturity process. Substantial improvements were found in four of the five major categories. Make was especially beneficial in reducing inventory and labor costs, but had considerable impact on customer service and lead time as well. Deliver showed a range of enhancements, but our analysis here indicated that these are minimal ranges of potential improvement. We expect more will develop as this part of the model continues to mature.

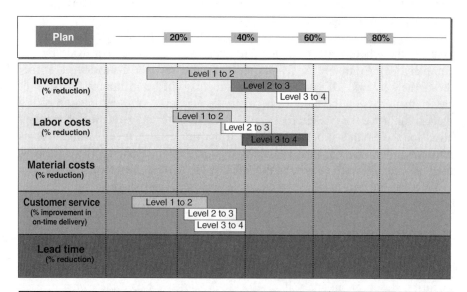

Figure 3.10. Moving Up the Supply Chain Maturity Model: How Much Value Is Created?

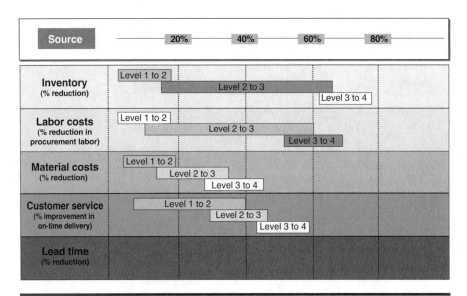

Figure 3.11. Moving Up the Supply Chain Maturity Model: How Much Value Is Created?

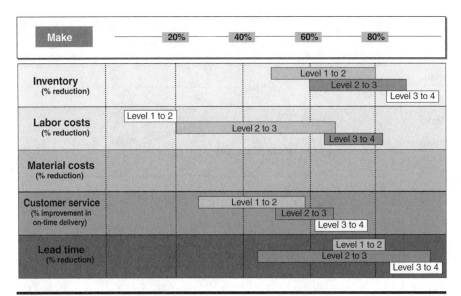

Figure 3.12. Moving Up the Supply Chain Maturity Model: How Much Value Is Created?

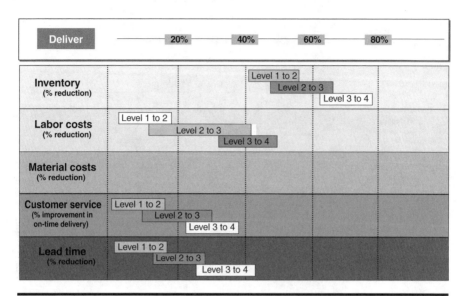

Figure 3.13. Moving Up the Supply Chain Maturity Model: How Much Value Is Created?

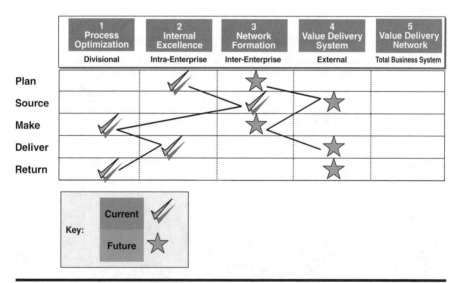

Figure 3.14. Moving Up the Supply Chain Maturity Model: How Do You Rate Your Own Company in Each Process? Where Would You Like to Be?

SETTING THE AGENDA: CURRENT AND DESIRED LEVELS OF PROCESS MATURITY

We now have all the tools we need to set the direction and priorities for improving the maturity of the business processes. Figure 3.14 shows a simple model used to illustrate the current and desired levels of process maturity. Current levels are the province of each company's management. The recent surveys done by CSC and *Supply Chain Management Review* suggest that few companies rate themselves much beyond level 3 in their current state of development. The assessment of the desired levels of process maturity is driven by a number of features, in particular:

- The value that can be driven out by investing in greater maturity
- Market pressures that demand levels of customer service
- Responsiveness that can only be achieved by more collaboration

Now we know where we are and the destination we want to achieve. Next we need some view of costs and time scales. The technology maturity matrices that were presented in Chapter 1 give part of the view of the likely costs and time scales to achieve the desired levels of maturity. The additional costs are those of the effort required to introduce the changes.

The process of analysis now enables us to establish a likely net contribution of value from the increasing maturity of each process based on:

- Value created, from the process maturity transition matrix
- The likely cost of delivering the supporting technology, from the technology maturity value matrix

With the rough order-of-magnitude costs and benefits, we can establish a supply chain process transformation program that is:

- Value based
- Prioritized
- Time based

THE IT DEPARTMENT NEEDS TO ASSUME A NEW AND IMPORTANT ROLE

To support this advanced level of effort, another important transformation must take place. The new view for members of the IT department is to move from being a cost center to being a strategic partner and help find and validate the improvements. They must also help manage IT investments so that there is a credible return on those investments. Finally, they need to help the firm respond to actual market demands through the most effective knowledge transfer system. This means that IT projects need to be aligned with business plans and goals and not be stand-alone efforts. Using business process management systems, IT must take the lead in developing the inter-enterprise knowledge sharing we have been discussing.

CONCLUSIONS

Within the framework we have presented, it is possible to assign responsibility for improvement and to track the enhancements directly to the P&L statement and the balance sheet. Doing so requires establishing the criteria of importance to be captured, roles and responsibilities for execution, and a tracking mechanism that verifies the benefits achieved. We have outlined a procedure for accomplishing these objectives. In the following chapters, we will further illustrate how actual benefits can be achieved through each level of the maturity model. Actual case studies, supplemented with the enabling technology, will be used to validate that what we have suggested is feasible.

4

GETTING BEYOND THE SUPPLY CHAIN ROADBLOCKS

This chapter details the changes an enterprise must make as it moves from a level 1 to level 2 position, and prepares itself for an advanced state of progress, and the values that can be generated. In an area occupied by few businesses, the enterprise gains control of its internal processes, segmenting the business into meaningful end-to-end organizational slices. Business processes are finely mapped, using state-of-the-art modeling tools, and the resultant models become instantly deployable in the business. Inhibitors to change are inevitable, however, and the major ones will be described, along with suggestions for overcoming them. Citing results from a recent survey conducted by CSC and *Supply Chain Management Review* magazine, the authors explain how and why most firms are bogged down in levels 2 and 3 of the five-level progression and what it takes to move into the advanced levels of progress.

GETTING BEYOND THE SUPPLY CHAIN ROADBLOCKS

With a determination to not only pursue supply chain excellence but also to reflect the results through meaningful financial improvements, the firm is prepared to further consider its progress through the five levels of the supply chain maturity model. Now we consider the changes that an enterprise must make as it moves from an entry-level position and begins to ready itself for the advanced levels of supply chain management. With adequate attention directed to the necessities of level 1, the firm can develop early savings, which can be traced to the financial statements and used to fund later efforts. The requirements in this part of the evolution are also essential for getting the firm ready for the use of business process management (BPM) and BPM systems, the tools of integration that will be applied across the value network that is eventually created.

Critical to this first level is the need for the business to gain control of its internal processes, so there is an adequate level of collaboration within the company's four walls. Then the total leverage of the business can be put to use to gain maximum business advantages across an extended enterprise. Getting the house in order so that the greatest overall network benefits are achieved is one of the supply chain absolutes, a necessity for best success. Any chance of optimizing the end-to-end business processes in the extended enterprise begins inside the firm. That is an extremely elusive concept for some business organizations.

Let us be clear about one point. Not every firm embraces supply chain management as a necessary business initiative, nor do some organizations and people accept the essential tenets of using collaboration and technology as the enablers to gaining the advanced positions. We have encountered many people in a number of industries where the existing business model and internal cultures are so steeped in lack of trust and fierce independence — between business units, functions, and parts of the hierarchy — that becoming prepared for supply chain management and using BPM to reach advanced levels, with the help of key external business allies, is virtually impossible. Content to apply time-honored practices, for example, which dramatically limit information transfer within the company and virtually exclude such transfer to external business allies, these companies and their leaders constantly work on local optimization efforts. Refusal to cooperate on projects that require the combining of resources and focusing on the firm's full leverage is another great inhibitor to progress. Such attitudes and positions become the major problems to the advancement we will discuss.

BEGIN WITH PROCESS CHANGES THAT ADD VALUE

As the firm does embark on its supply chain management journey, it invariably initiates a search for immediate improvement to areas that offer financial gain. To assist this effort, process maps are drawn detailing the important process for whatever is defined as the end-to-end business. Most of these maps are developed on an individual business unit basis, which is very acceptable for early

Approach	Business Case Focus	Key Considerations	High-Level Plan
Integrate and improve key business processes first	Improving business process efficiency & effectiveness	Must take a process view of running a business	Process redesign / Process automation

Figure 4.1. Approach to BPM in Supply Chain Management: Level 1 to Level 2

efforts. Using current modeling tools, as described by Poirier (2004), the firm can move forward to implementing the enhancements in the areas of greatest importance. The early gains confirm the potential of a supply chain effort, but eventually lead to the need for cross-organizational collaboration.

This effort generally gets off to a good start, because of the potential benefits that can be achieved. The secret is to make certain the major cultural inhibitors are overcome, so the progress can continue to higher levels. As mentioned, the survey conducted by CSC and *Supply Chain Management Review* verifies that most firms do make early gains and can document financial gains, but get bogged down in the middle of the effort. As the recommended steps are discussed, we will take care to point out the more salient inhibitors we have encountered and the solutions that aid progress.

The general business approach taken is described in Figure 4.1. There we see that the firm analyzes its business processes and begins an attempt to improve those that have the greatest opportunity for short-term improvement and financial impact. Building a business case for making such an effort is done next as the firm attempts to improve both efficiency and effectiveness across the highlighted processes. The important considerations must be oriented around taking a process view of running the business, a step often missed in early efforts. With a high-level plan established, the firm can then begin working on process redesign and process automation, where appropriate.

Starting on the inside requires some basic understanding. To begin, the need for collaboration, or the ability to work jointly with others or together, especially in an intellectual effort for the greater benefit of the overall business, is crucial. Supply chain progress is not lacking because of an absence of tools, brains, or effort. It lags in many firms because of a lack of cooperation between parties that have the ability to help each other. Our advice is to begin by selecting one of the more progressive business units or functions within the business and build successful implementations, requiring some form of intra-enterprise cooperation. Then the firm can replicate the effort across more units and functions and proceed up the maturity model.

FOLLOW THE THREE A'S APPROACH

As the focus intensifies on internal improvement or the movement from a beginning state through level 1, there are several basics that require attention. We refer to these basics as the "three A's" — awareness, alignment, and action. To begin, at the business level, some form of business audit or diagnostic is appropriate to establish the opening position vis-à-vis the industry, best external practices, or reliable benchmarking data and an idea of where the firm wants to go with its improvement effort. With this information as background, an *awareness* session should be conducted with the senior managers, to raise their understanding of the potential value of supply chain as an improvement tool, the importance of following a framework such as the SCOR® model or our maturity model, and the financial improvement that can be added to their areas of responsibility by progressing to higher levels of the maturity model. It is a fatal mistake in supply chain efforts to assume that all senior managers understand or are aware of the concepts, intentions, and necessary involvement behind such efforts. The awareness session is extremely important to make certain those leading the organization understand exactly what they are committing themselves and the resources under their direction to accomplish.

The second important deliverable in this part of the effort will be gaining the necessary *alignment* among these leaders around the supply chain vision and the guiding framework. Expect that as this alignment is being sought, most managers will be delivering endorsement for the effort, some of which is genuine and some of which is cosmetic. We find during these early stages that business unit and functional leaders generally can be broken down into two classes: advocates and cynics. The *advocates* sense the potential advantages and quickly endorse the supply chain effort. The *cynics* withhold actual endorsement and implementation activity until they see results that verify the possibilities for their areas of responsibility. A second session or part of the effort dedicated to gaining overall alignment to the necessary framework and changes that will occur is very important to eventual success.

When awareness of the values to be added through a concerted supply chain effort is realized, and there is alignment around the framework, action plans, and required business transformation, then we find most organizations are prepared for *action*, and the necessary implementations begin to occur. These simple steps have proven to be very valuable for those businesses making the transition to higher levels of supply chain progress.

MAKE THE TRANSITION THROUGH
FROM LEVEL 1 TO LEVEL 2

With commitment to advancement and greater success, the firm is then ready
for the establishment of a supply chain strategy, a process improvement program, and a means of ensuring project management over the many initiatives
that invariably appear. Now the firm goes through an analysis of the processes
involved in the highest potential areas, to isolate specific process improvements
that have the greatest value — initially for the internal firm. An organizational
design follows this step, as someone must be selected to lead the supply chain
effort and a minimal staff put in place to help with the necessary intra-enterprise
initiatives.

Figure 4.2 can be used as a guide to making the transition from level 1 to
level 2. It describes how the as-is state in a critical process area is defined and
digitized, so an improvement effort can be launched and completed. The hy-
pothesis behind this approach is that the eventual value chain that is constructed
must move from one enterprise driving many process improvements for internal
benefit to common enhanced processes driving mutual benefits for many en-

- Digitizing the "as is" – helped by the new BPM applications — bringing immediate benefits:
  — Reduced lag time and duplication of effort
  — Actual operations analyzed and measured, yielding improvement opportunities
  — Improved processes through redesign
  — Increased control and monitoring
  — Business rules enforced for optimum operating conditions

- P&L benefits:
  — Increased productivity through task management:
    • Less error-prone process execution
    • Faster process deployment
    • Reduced order processing costs
    • Fewer mistakes, expedited orders, emergency freight costs
  — Manufacturing labor costs down through less disrupted manufacturing
  — Reduced material costs:
    • Lower wastage
    • Better scheduling of inbound freight
  — Reduced information technology costs:
    • Avoid replacing or upgrading legacy systems

- Balance sheet benefits:
  — Reduced inventory

- Key performance indicators:
  — Shorter and more accurate lead times as planning is synchronized with order promising
  — Delivery available-to-promise during order taking

Figure 4.2.   Transition from Level 1 to 2: The Benefit Case

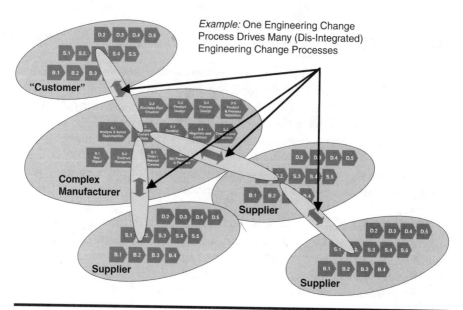

Figure 4.3. Collaboration Across the Value Chain: One Enterprise Launches a "Cascade" of Processes to Suppliers

terprises. As the figure explains, the firm begins to digitize the current conditions in key processes, especially those that can be enhanced with BPM tools. With this information, the profit and loss benefits are isolated in terms meaningful to the firm, such as increased productivity, lower manufacturing costs, less material usage or waste, and a reduction of information technology (IT) costs (or other factors pertinent to the particular business). Balance sheet items also receive attention, including any potential reductions to inventory and carrying costs. An output from this effort should also include any key performance indicators that serve to measure and document the improvements being made. The essence of the effort is to construct a benefit case for moving forward, with at least some order-of-magnitude understanding of the values that can be added.

Figure 4.3 is a visualization of the state we are proposing, where business constituents, internally and externally, are collaborating to make, for example, one engineering change to a process, which drives many other valuable engineering changes to affected processes. The collaboration starts with internal cooperation between functions and departments, but it quickly escalates to getting the help of many members of the extended enterprise in order to come close to optimized conditions. This technique begins with an awareness of what is being proposed. We define *collaboration* as working jointly in a business environment with other parties in an intellectual manner, intended to provide greater knowledge and understanding for the participants, so they might find optimized operating conditions across the total enterprise. Bear in mind that collaboration used to be as simple as walking around a corner with the help of a guide. Now the corner could be anywhere in the world, and in today's complex business environments, there are plenty of new corners.

THERE ARE SIX DOMAINS OF CHANGE

In essence, the move from level 1 to level 2 becomes a move from a functional organization to a process-based organization, with all the advantages that brings — shared performance targets, free flow of information, reduced lead times and inventory, and better customer service. There are six aspects inherent in such a change:

- **Process changes** — Redesigns or reengineering of those processes that are deemed to have importance for a particular business unit, and especially those processes that could add value for multiple parts of the firm if improved.
- **Organizational changes** — Establishment of a serious supply chain group, which will guide further progress, and the connections necessary with enabling technology (the link with the IT group) and members of the business units expecting added value from the supply chain effort.
- **Changes to the location of work** — Necessary internal modifications so operations can be brought as close to optimized conditions as possible. Elimination of manual processing in favor of error-free electronic transmissions will facilitate such changes.
- **Changes to the data needed to support the new ways of working** — This effort begins with identification of what information/knowledge is essential to the improvement effort and moves to alterations that speed the transmission to the point of need, with impeccable accuracy.
- **New IT applications** — A critical step that must apply technology as an enabler to the improved processes and not revolve around selection of software packages that have nebulous value across the organization.
- **New technologies** — Consideration of the values to be gained and the application of BPM tools are especially important in this domain, as the firm decides how to collaborate internally and externally to add value across the end-to-end processes and select the most appropriate technology for itself and its closest business allies.

In any reorganization, these domains of change must be considered, and those responsible for the process of change must take them into account. Moving from level 1 to level 2 will undoubtedly require changes to processes and organization and involve changes to other domains as well.

SEGMENTING THE ORGANIZATION INTO ITS HORIZONTAL COMPONENTS

So how does this phase of the effort progress? Step 1 is to identify the horizontal components of the organization, using a technique called supply chain mapping. By mapping the product flows and assets used, self-contained product families can be identified. These families will form the basis of the new organization structure.

Figure 4.4 shows 1 of 20 maps drawn by Clarks Shoes, one of England's largest shoe manufacturers, during an exercise to reengineer its global supply chain operations. Each map proved to be a self-contained business within the overall business. The products in each map used a unique set of assets and thus could be managed in isolation from other product families. In this way, the firm identified the organizational components of its new business model.

With the elements of the new organizational structure identified, the business can establish the key metrics that govern each product family — a mixture of financial and nonfinancial measures. Nonfinancial measures include such elements as:

- Lead time
- On-time delivery and fill rate
- Customer service measurements
- Sales forecast accuracy
- Stock record accuracy

The financial measures are, of course, the profit and loss account and balance sheet for the product family, together with the return on net assets (RONA) calculation that becomes possible. Figure 4.5 shows the product family "dashboard" that became the governing reference for changes at Clarks. In this case, the firm decided to segment its business by sectors and major product lines. The key performance indicators were then arrayed against each major sector so that performance could be easily monitored across the internal organization. Similar dashboards could be constructed for other firms in other industries with the pertinent metrics.

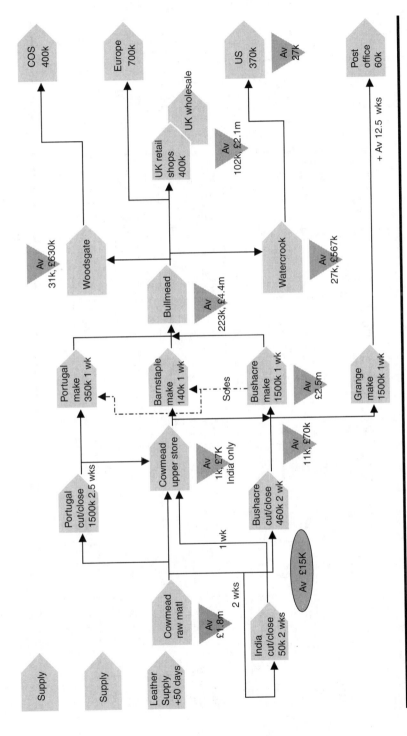

Figure 4.4. The Desert Boot Supply Chain Map for Clarks

	Men's Far East	Canvas Far East	Women's Brazil every day casuals	Men's formal Italy	Children's casuals Europe	Bushacre factory desert boot/Men's premium casuals	Rush Hill factory	Abbeyfield factory	Millom canvas (not slippers)	Springer factory	●	○
Stock to Sales Ratio											< 25	> 35
Margin											> 5	< 3
Turnover											> 20	< 10
RONA											> 60	< 40
Lead Time (Longest)											< 8	> 20
Lead Time (Average)											< 8	> 15
Delivery Perf (to w/h)											> 95	< 85
Delivery Perf (to cust)											< 95	> 85
Markdowns Retail											< 5	> 10
Markdowns Wholesale											< 2	> 10
Returns Worn											< 0.5	> 2
Returns Unworn											< 0.1	> 0.5
Forecast Accuracy											< 10	> 30
Stock Accuracy											> 95	< 90

Figure 4.5. Product Family "Dashboard" for Clarks

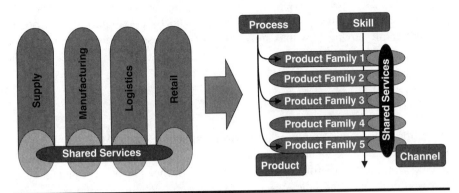

Figure 4.6. The Organizational Changes

Another essential organizational change in moving from level 1 to level 2 is the breaking down of the old functional silos, and creation of new horizontal structures, with the ability to manage all aspects of the product family: range selection, sourcing, manufacture, storage, delivery, and sales. Figure 4.6 depicts the kind of transition being suggested. To effect real collaboration within a business organization, the best practices attained by various business units or functions must be shared across the firm, resulting in at least optimized internal processing. The four key roles in the new horizontal organization are:

- **Product family owner** — A determination of who holds responsibility for the processes associated with each particular product within the portfolio
- **Process owner** — A further determination of who is responsible for the key process steps across the end-to-end value chain associated with the individual products
- **Skill owner** — Identification of those people who possess the necessary skills to reengineer the processes involved and collaborate with other parties to bring the enhanced processes as close to optimization as possible
- **Channel to market owner** — An often overlooked role, in which the people necessary to assure the efficient delivery of products and services are identified and their channels to market explicitly identified, so that optimized conditions can again be pursued

The key role in this new business order is product family owner. This is the person(s) responsible to the enterprise for maximizing the RONA in the product family — by deploying the best processes and skills (the old functions now become servants to the process organization) and by collaborating with other product family owners to maximize the effective use of shared channels to market.

STARTING THE PROCESS OF CHANGE

With this background awareness, the firm can start to design the way it wants the business to work — the new processes and supporting information technology — with a set of metrics to judge the impact of the changes, not just on each function but across the whole product family supply chain. No two companies approach these changes in the same way, but the effects of moving from level 1 to level 2 are marked and produce similar results. We will look at each of the major processes in turn, beginning with Plan.

PLAN

The Plan process is perhaps the one that shows the most marked change from level 1 to level 2. This is the first step along the integrated supply chain, as functions begin to plan their activities jointly.

Figure 4.7 reminds us of the change in maturity. Specific functions and business units, with few integrated processes, carry out demand and supply planning at level 1 internally. Production plans and inventory plans, for example, would be carried out plant by plant — based on local forecasts of demand and buffered at each stage by inventory. Often, purchasing decisions would be taken on the basis of local purchasing performance measures (least

Level 1: Demand/supply planning is done internally, with no integrated processes and tools across plants	Level 2: Global demand/supply planning is consistently aggregated across the firm, focused on functional accountability, & continuously improved by comparisons to historical performance

Figure 4.7. Plan Process Maturity

unit cost, for example), with little consideration of the impact of that decision on the overall performance of the internal supply chain.

Level 2 shows a marked difference. Now the upstream activities are planned against the same demand forecasts as those produced through the downstream-side, customer-facing activities. Inventory is planned strategically across the whole chain, balancing holding costs against usability and lead time. Ideally, inventory would be held as far upstream as possible, because that makes it most flexible in its end use, and as far downstream as possible to meet customer lead time targets. Somewhere in the middle is a balance point at which RONA is maximized.

CASE ILLUSTRATION: ZENECA AGROCHEMICALS

To illustrate the possibilities, consider the example of Zeneca Agrochemicals, now part of Syngenta. After an exercise that identified 22 major self-contained product families, the insecticide/fungicide suspended colloids family was taken as a pilot. This product family is comprised of 5 active chemicals which, in turn, are made into 20 different formulations (different insecticides and fungicides) and ultimately 100 different market-pack combinations for sale to farmers in 35 countries around the world. The product family was manufactured in two stages: (1) active ingredient manufacture and formulation and (2) filling and packing. Manufacture took place in two U.K. and one European plant, with active manufacture near Edinburgh and formulation, filling, and packing in the southeast U.K. and in Belgium. Each plant was traditionally planned based on local performance measures of plant utilization. Inventory of active ingredients, once created upstream, was shipped to the downstream plant and therefore had no adverse effect on the upstream plant's measures.

The initial product family study gave the improvement team an end-to-end supply chain map similar to the Clarks example. The other tool the team devised was a simple spreadsheet model showing the RONA in the chain, as illustrated in Figure 4.8. This was both a performance measurement tool to determine if RONA was going up or down and an enablement tool that helped to set priorities between some of the improvement options. For example, the team could evaluate different batch size rules and resultant changeover costs and stock levels. The RONA model was used to determine which choices led to maximum RONA.

This new approach led to an initial view of the profitability of the products in the overall supply chain relative to the assets that they actually consume. It was, of course, not the only performance measure used, but it was certainly a key one and one of the most innovative.

Sales	100
Gross margin	70
Less direct costs	(20)
Less indirect costs	(15)
RONA contribution	35
Fixed assets	30
Stocks	50
Debtors	35
Creditors	(25)
Assets employed	90
RONA %	39%

Figure 4.8. Zeneca Case Study: RONA

The other measures used included:

■ Delivery due date performance — The difference between actual and promised dates

■ Availability from the different stock points — The reliability of shipping from each source

■ Adherence to manufacturing schedules, and the percentage of time that schedules were kept

The improvement process began with an agreement on customer service targets for the end of the chain. These targets were for each market and for each type of product and customer within the market. To set goals for the supply chain, a customer service mission was needed, which was:

■ **Specific** — Precision about which products will be available from stock and which can be manufactured to order

■ **Measurable** — Targets, which customers will accept, for the availability of products from stock and the lead time for making to order

■ **Seasonal** — For this type of very seasonal product, different mission statements were needed for the period of peak sales and the dead season

The customer service mission was then used to drive the activities of each stage of the supply chain. This part of the effort began by looking at the way in which the downstream plant fed products to the selling companies. By going back to the basic sales data, the plants' ability to fulfill the end customer needs was assessed. This was different than the batched requirements that the plants would normally see coming through from the selling companies, driven by the existing distribution resource planning system. For each major part of each plant, the team established:

- Its capacity, based on labor and machine availability and manufacturing capacities
- The load imposed on the plant by the base sales, taking into account batch sizes, changeover times, and target lead times

The picture that was painted, however, did not seem to match the full plant conditions that the factories were experiencing. The reason for this was the fact that the planning process currently in place was being driven by people who believed their job was to fill the plant with orders. The conclusions reached from this work were that:

- Plant capacity could match demand from the selling companies within the week that those sales will be made, with no exceptions.
- The selling companies could work on reduced lead times and small buffer stocks.
- There was no need for a seasonal stock build ahead of the selling companies' peak selling seasons.

The first large impact of planning across plants was a major revelation about ability to serve the customer with far less inventory than had ever been achieved before. This was measured on the RONA model, so that the financial impact was plain for all to see. Figure 4.9 shows the impact on RONA of reducing

	Baseline	With inventory reduction
Sales	100	100
Gross margin	70	70
Less direct costs	(20)	(13)
Less indirect costs	(15)	(15)
RONA contribution	35	42
Fixed assets	30	30
Stocks	50	44
Debtors	35	35
Creditors	(25)	(25)
Total assets employed	90	84
RONA %	39%	50%

Figure 4.9. Using the RONA Model

inventory. Not only did inventory fall, but direct costs incurred also fell, despite a small increase in changeover costs in the active plant — more than compensated for by lower storage, waste, and financing costs.

The next piece of work looked at the upstream plant and the way it fed the downstream plant. As is common in the chemicals and pharmaceuticals industries, capital expenditure had been focused over a number of years on matching the available capacity with the total requirements in the year. This balance between fixed capital and working capital is not necessarily accomplished with optimal results. The team looked at the balance again, using sales at the end of the supply chain instead of batched demand.

The upstream processes were de-bottlenecked, by managing changeovers in a more effective way. This change entailed some capital investment. It allowed the team to move away from steady manufacture, in large batches sometimes running for many weeks, toward short cycle manufacture, responsive to demand in the rest of the chain. This change minimized the necessary stock build.

The team came up with preconceived ideas about what process changes would and would not be allowed. The most significant of these concepts was the belief that the board would not approve the capital expenditure needed to de-bottleneck part of the active ingredient plant, at a cost of perhaps £500,000. This disbelief was put aside by using the RONA model, which showed that the payback from the investment would be cash positive throughout the effort and would also have a positive impact on both the inventory and RONA in the supply chain.

Inventory could now be managed across the whole supply chain, taking account of the capacities and lead times throughout the linked processes. The right levels of raw materials, upstream intermediate materials, and finished goods could be held in the right balance and adjusted as the seasonal demand changed. The overall level was reduced dramatically, resulting in a much leaner, more responsive supply chain. Local initiatives were put into the context of the whole chain to avoid double-counting stock and capacity buffers. Once again, planning across the stages in the internal supply chain made a dramatic difference in the way the operations were perceived, with a huge positive impact on RONA for the product family.

OTHER IMPACTS FROM THE CHANGE DOMAINS

Returning to the six domains of change, this exercise resulted in changes in all six! The business processes were changed fundamentally, as was the organization and the way its performance was measured. The organization now had a product family owner, and reporting lines were changed to reflect this step. The

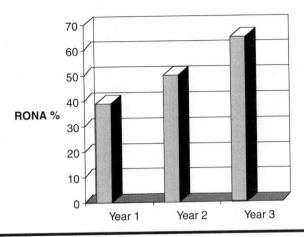

Figure 4.10. Performance Measurement

major players in the product family had performance goals linked to the overall chain, not just their local plant. Location was affected, because decisions would be made for the whole chain, instead of locally at any one plant. Finally, the data, applications, and technology used to manage the product family changed, as the learning was built into the implementation of SAP R/3. Most importantly, the RONA for the whole chain increased dramatically, as shown in Figure 4.10.

OVERCOMING THE OBSTACLES

So what gets in the way of planning across the supply chain in this manner? The first major issue is the existing organization and its traditional cultural imperatives. Reporting lines into the old functional organization get in the way of pan-supply-chain thinking. Putting together a change team, with a group target that cuts across all existing reporting lines, requires careful selling to the existing management, to achieve alignment toward execution. At Zeneca, at no time during the exercise did that present any problem at all. Everyone was kept fully aware of what we were doing and what the team's objectives were, and this was sufficient to free up the operation.

The second issue is that most of the information in the level 1 business is held by function. There is no supply chain information collected together to give an overview of what is happening to this whole piece of the business. Supply chain maps help to overcome this, presenting information in an organized way that people can readily understand, challenge, and work with. The initial reaction to a supply chain map is often disbelief. Never having seen a picture of an entire supply chain laid out like this, there is often a reluctance to accept that there is that much inventory and the lead times are as long as the map indicates!

Often, the first stage of the process of improvement is to go back into all the numbers represented on the map and confirm that they really are accurate. The more they are poked, the more robust the numbers prove to be — and so we move from disbelief to acceptance! The more there is acceptance, the greater the alignment behind the necessary process changes and business transformation will be.

SOURCE

Moving Source from level 1 to level 2 is a similar exercise in pan-supply-chain thinking. Again, the key is to run the Source process for the good of the whole chain, not just the local performance of the purchasing department or a particular business unit. Level 1 sourcing is characterized by such issues as a tactical focus on price, fragmented spending, a decentralized procurement organization, lack of knowledge for spend categorization, and minimal authority within the procurement organization.

With a pan-supply-chain view, Source can begin to manage commodities across the enterprise, unlocking economies of scale. This is often coupled with a rationalization of the supplier base, so that more time can be devoted to negotiations with individual suppliers, already beginning to look beyond level 2, to level 3 collaboration. The characteristics now change toward a focus on total cost, the organization becoming a hybrid decentralized/centralized model, an increasing knowledge of spend categorization throughout the organization, and a move toward achieving some level of authority within the procurement organization.

Boeing, following its acquisition of Rockwell and McDonnell Douglas, formed a shared services group to manage $3.5 billion of nonproduction purchases. A global e-procurement system to establish compliance was a feature of this effort. Airbus is another example from the world of aerospace. Airbus was created from a consortium of companies from four European countries. In 2004, it produced and delivered more commercial aircraft than any other company, including Boeing. It is currently developing the world's largest passenger aircraft, the A380, due in service in 2006. In order to maintain its current competitor advantage, it targeted a significant reduction in the total cost of production. Key to achieving this intention is the effectiveness with which it manages its suppliers. The Airbus Sup@irworld program and a single supplier database and set of interfaces are a key part of this effort. In particular, Airbus now has consolidated spend reporting and management across all the operating areas and geographies. This includes a single supplier quality signoff, a single approach to monitoring supplier performance, and a consistent set of contract terms. When the scale and complexity of Airbus are considered, this is a massive achievement, in moving from a much-dispersed level 1 to a well-consolidated level 2, and will undoubtedly result in lower cost and better service from the supply base.

MAKE, DELIVER, AND RETURN

A similar story unfolds as we look at the other key processes, as they move from level 1 to level 2. In each case, we see the traditional isolation of functional organization, driven by local performance measurement, giving way to activities planned and executed in the full recognition of the impact they have on the rest of the enterprise. This often benefits the traditional function more than people would admit, as they are asked to give up what they often see as the "right way to do things" or "the way it's always been done."

One of the most powerful measures in manufacturing has been utilization — whether manpower or machine. It has taken decades and the concerted efforts of many leading thinkers to replace this metric with more effective measures such as schedule adherence. Of course, many now see the fallacy of working assets even when there is no demand for their output — it just results in inventory sitting in the system. But the old logic is still seductive!

Take the example of a U.K.-based automotive components manufacturer. It had installed a new material requirements planning (MRP) system and processes to provide sales forecasts for procurement and production. Success was monitored by the level of finished goods inventory and on-time dispatches. Unfortunately, inventory was higher than it had ever been, and the dispatch success rate was falling behind dramatically. What went wrong? Analysis soon revealed that the business had two distinct components — one that provided a service to in-country customers, with regular, relatively small orders, and another that provided infrequent, large replenishment orders to overseas agents. As these order streams were for the same products, the sales forecasts, provided in good faith by the sales department, always overestimated the regular demand — causing stock levels to rise. But, of course, there was never enough stock to cover one of the larger orders — so stock-outs resulted, the regular customers were starved of stock, and the overseas order was late.

The underlying cause of the problem was level 1 thinking — sales providing the forecast it thought manufacturing wanted, manufacturing trying to second-guess the actual demand and playing catch-up to sort out the delivery shortages at the last minute, and the warehouse showing poor dispatch performance at the end of the internal supply chain.

Moving the whole situation to level 2 produced the right results and required changes in the way processes worked across the three functions involved. The company carried out a detailed root cause analysis of stock, the pattern of recent sales, and the processes employed. The two separate demand patterns were recognized through a reclassification of products and customer service goals. Manufacturing lead times and minimum batch sizes were rebalanced to suit the market demands that were now recognized. Most importantly, the business processes for forecasting, production planning, and dispatch were redesigned and the effects reflected in a reimplementation of the MRP planning tools. The company realized an increase in product availability in one of its two major product groups from 62% to 96% with no increase in stock and in the second an increase from 86% to 96% while reducing stock by 34%. This situation provides one more impressive result for a change in thinking from level 1 to level 2.

MANAGE PRODUCT

The final process areas to consider in the context of the transition from level 1 to level 2 are the Manage Product processes. In the same way as the supply chain at level 1 is managed functionally, so are the management of the product portfolio and the design and introduction of new and modified products run in silos. The design tools and product data are typically the province of the engineering department. There is little organized discussion between functions about the merits and requirements of a modification, so sales forecasts become unreliable, manufacturing is always a revision behind and hit with increasing levels of rework, and the sales department spends much of its time apologizing.

Maybe this condition is a little overstated, but when the processes work to maximum, the impact on performance is significant. In a recent project with British Aerospace to look at the management of engineering changes for the Typhoon Eurofighter, the reorganization of the business processes for change management resulted in an 80% reduction in the cycle time for manufacturing-required design changes and 50% for product design changes. Data errors were reduced by more than 80%.

The co-location of multidiscipline teams in design studios, at companies like Thales, RACAL, and Strachen & Henshaw, has seen significant paybacks: project budgets reduced by 20%, material costs reduced by 10%, product changes reduced by 50%, management overhead reduced by 50%, and increased productivity leading to up to 15% less staff required. The changes have all been made within the walls of the organization — combining the knowledge of design engineers, production experts, customer-savvy salespeople, and procurement specialists — to create a powerful internal set of processes, which benefits not only the company but also its customers and suppliers.

CONCLUSIONS

Companies are now adopting the ideas behind making supply progress more freely — but there is still a long way to go. The successes are still outweighed for many by the inhibitors of traditional thinking:

- **Functionally oriented organizations** — So that problems rarely can be tackled at root cause, which often may be elsewhere in the organization
- **Local performance measures** — Which reinforce the functional organization: "I'm doing okay; my performance measures tell me so!"
- **Lack of effective process thinking** — So that processes stop at the functional boundaries, adding time and cost

Results from a recent survey conducted by CSC and *Supply Chain Management Review* magazine bear this out. Most companies at level 1 do not leverage scale across the entire enterprise. Content to find savings on a functional or business-unit basis, they adopt a stovepipe mentality that sees no advantage in centralizing any function or sharing any supply chain improvements. Collaboration between functions or business units is resisted. Communication systems to facilitate processing throughout the organization are nonexistent. And, of course, without seeing the true benefits of moving away from level 1, the investment in time and people is too much — so these companies stay where they are, and sweat the functional assets and struggle in a world where others are moving on.

There are two saving graces. First, the authors have put a range of values on the transition from level 1 to level 2. In a survey of over 40 case studies, the results of which are summarized in Figure 4.11, significant gains were recorded through process improvements in the five SCOR® processes. They were measured against five key metrics:

- **Inventory** — Impressive reductions of up to 85% in inventory levels were made, with the major process driver being Plan, where 50 to 80% of inventory was eliminated.
- **Labor costs** — Again, Plan was the major process to contribute labor reductions of up to 35% in the processes concerned.

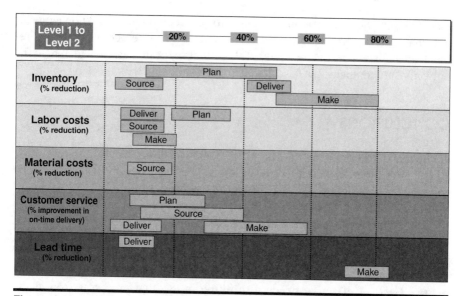

Figure 4.11. Moving Up the Supply Chain Maturity Model from Level 1 to Level 2

- **Material costs** — Taking Source to level 2 unlocked between 5 and 10% of material cost.
- **Customer service** — The first of two nonfinancial measures, it was improved by as much as 60%.
- **Lead time** — The second nonfinancial measure, it saw gains of up to 80%, through the Make and Deliver processes.

So the true benefits of increasing process maturity are measurable and can and are being used to justify the investments to move from level 1 to level 2. Processes can now be finely mapped, using state-of-the-art modeling tools, and the resultant models become instantly deployable in the business. It is no longer a problem to make the process changes necessary to achieve level 2. We will expand on this as our story unfolds through the higher levels in the following chapters.

5

MOVING EFFECTIVELY TO ADVANCED LEVELS OF PROGRESS

As the business firm overcomes the pitfalls encountered in the early levels of its supply chain progress and has its internal house in order, in terms of its ability to develop and share best processing electronically and on an intra-enterprise basis, the transition to higher levels can occur. Now the firm moves, with the help of a few carefully selected business allies, into an extended enterprise or network environment, in search of higher orders of improvement, which can bring additional benefits/values to all participants. Mattel, Inc. found higher values in its network, when that toy maker used the Internet with the help of designers and licensees to collaborate on new product design. The firm moved design on-line so that virtual models of new products could be transferred electronically, instead of through the usual manual system. Development time was reduced by 20%. By digitizing and automating the transfer of information between Mattel and its licensees, the company found that approvals could be accomplished in 5 five weeks, instead of the normal 14.

Unfortunately, the idea of network collaboration does not come easily for most organizations, nor does the technical means to facilitate such cooperation. Indeed, the concept of creating greater value through an intelligent value network can be elusive. In today's business environment, selling value is an extremely difficult job, as there are few people who understand the idea or will expend resources to develop these values. Those willing to do so generally reside at high levels in a few firms and are opposed by an army of others with an incentive to continuously cut costs and save money, generally at the expense of willing suppliers. It takes a strong-willed business leader, and supportive supply chain managers with the help of information technology, to forge ahead in the face of such obstacles. Lockheed Martin pushed forward in this area and linked 80 of its major global suppliers into a network for designing and building a new stealth fighter plane. As part of a $225 billion project, the firm is linking a global network with real-time access so the key suppliers can work simultaneously with other global partners. The expected savings could reach $250 million over the ten-year life of the project.

COLLABORATION IS BEGINNING TO DEMONSTRATE ITS VALUE

We can see that in spite of the reluctance to work with external business partners, collaboration can and does occur, as the firm moves through its maturity model. Figure 5.1 corresponds to the five levels of that model and describes the usual transition that takes place. Beginning in the *inform* level, there are a series of one-way communications set up, primarily to handle design changes, to enter data through an electronic data interchange system, and to process electronic funds transfer. In the *interact* level, the firm breaks down its internal resistance to cooperation and begins to carefully exchange data important to optimizing across the full organization with a select group of business allies, often on a pilot or experimental basis.

As the chasm to level 3 is crossed, the firm enters the *transact* area and begins conducting business with some of its external partners on an electronic basis, usually by linking some part of its planning system with selected business allies. Further progress becomes difficult, for the reasons cited, particularly the reluctance to share information externally. Fortunately, some firms prove the value by making customers, suppliers, and distributors a part of their supply chain effort, providing services to each other as another means of reaching optimum conditions across the end-to-end network that forms. In the fifth level, a *community* environment is achieved, where the linked firms fully utilize a collaborative business/operating model for interaction, with elements of customization, personalization, and community building appearing.

While many firms resist this type of progression, a growing body of evidence shows that more value is generated as enterprises overcome their normal proclivity not to collaborate and do share vital information outside of the four walls defining the firm's business. Examples where business allies, such as Tesco and Nestlé or Procter & Gamble and Wal-Mart, share database information on consumer trends and current market conditions to improve the results from special sales events form one such basis.

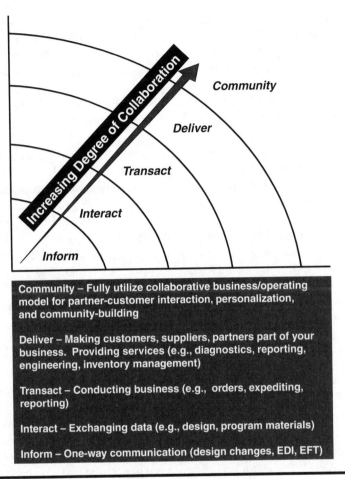

Figure 5.1. Understanding Levels of Collaboration

CASE EXAMPLES ILLUSTRATE THE POTENTIAL

BMW, the German automaker, offers a specific case example. This firm has been hard at work on a project to move part of its manufacturing to a "build-to-order" or customized system. The idea is to make most cars in Europe in that fashion, and up to 30% of the cars sold in the United States, without incurring extra cost. The secret? Linking dealerships (orders) with the factories (manufacture) and suppliers (raw material and parts). Customers are provided more options, with a third of the normal delivery time, while the company slashes the cost of overstocking. "Instead of choosing from a pool of dealer-purchased cars, buyers now can design their own Bimmer — from 350 model variations, 90 exterior colors, and 170 interior trims. To give customers this luxury, BMW overhauled its entire network, from sales systems to distribution software." After a customer selects his or her options, the dealer enters the order and receives a precise delivery date in five seconds. The data on the car are relayed to suppliers that ship the components in sequence. "The cars arrive 11 to 12 days later, one-third the time it took before the online system was in place" (Edmundson, 2003, p. 94).

Boeing's announcement that Japanese suppliers will design more than one-third of its 7E7 airplane offers another example. In this case, Boeing is departing from its traditional approach, in which the company carefully guarded its techniques for designing and mass-producing commercial jetliners. The idea was driven by Boeing's desire to cut construction costs and the time to bring the plane into use. Under the new scheme, external business partners will not only design a substantial part of the new model, but will build up to 65% of the 7E7. The business allies chosen include Mitsubishi, Kawasaki, and Fuji in Japan; Alenia Aircraft of Italy; and Vought Aircraft Industries of Dallas, Texas. Central to these moves is the new Boeing objective of "moving as much costs and risks off the books as it can," according to Richard Aboulafia, aerospace analyst at the Teal Group (Acohido, 2003, p. 2B).

Other examples span a variety of industries. As Tony Kontzer reports in *Information Week* (2003), "No one needs to sell Tim Swan on the benefits of collaboration. As global IT product manager for Barclays Global Investors, the asset-management unit of Barclays plc, Swan has deployed team-collaboration software to speed everything from contract management to deal flow." When the firm responds to customer requests for proposals, employees are able to "access previous proposals as well as set up online workspaces and tap into subject experts." The result is faster proposal generation and a higher possibility of gaining new business.

Using electronic collaboration is common practice within the four walls of Barclays, and Swan intends to extend its use across the firm's external network, into the area he calls "the virtual space between [Barclays'] external and internal firewalls." The model being employed is one that blends applications ranging from instant messaging to knowledge management, to introduce a "new breed of collaboration software and a new level of communication." The idea is to allow real-time interaction for employees handling information that can change in a matter of minutes. The most sophisticated companies, Kontzer reports, "want to imbed presence awareness — the ability to detect the online status of others on the network — throughout their networks, so employees can find available experts without looking away from their screens" (Kontzer, 2003, pp. 29–30).

Whirlpool found the value of collaboration when it commercialized a series of innovative products — appliances and storage systems for garages, dubbed the Gladiator line. To meet stringent standards set on cost and time to commercialization, the Whirlpool team charged with the development and introduction bypassed the traditional internal design and manufacturing system and outsourced the manufacturing of everything except the appliances. The Gladiator team worked with suppliers that were new to the company, in one case sourcing "tooling from a supplier that delivered it at one-third the cost and one-third the time of the company's current suppliers. Similarly, the team utilized the design capabilities of several suppliers in order to save time" (Andrew and Sirkin, 2003, p. 83).

RESULTS ARE SIGNIFICANT

The careful sharing of data and knowledge, previously considered proprietary or even sacrosanct, opens the way for the firm to work its relationships with the most important and immediate supply chain neighbors — customers, distributors, and suppliers — to find other hidden values within the extended enterprise. Organizations willing to take the necessary steps with selected business allies can reasonably expect to achieve the following results:

- Shorter lead times and cycle times, often reduced by as much as 40 to 50%
- Better, more accurate order entry and tracking, requiring far less reconciliation
- On-line visibility of raw materials, work in process, and finished goods across the end-to-end supply chain — with the ability to divert goods in transit
- Less need for inventory and safety stocks, coupled with an increase in inventory turns by as much as four to five times
- Lower warehousing and transportation costs, resulting in a 5 to 10% reduction in freight bills
- New revenue opportunities, often in nontraditional areas
- A reduction in general, selling, and administrative costs in the range of 5 to 10%

Using business process management (BPM) technology to facilitate this careful sharing and to find the savings cited introduces a new avenue for process improvement, typically at the point where most businesses are faced with diminishing returns from their supply chain efforts.

Approach	Business Case Focus	Key Considerations	High-Level Plan
Improve processes and infrastructure in parallel	Business transformation	Integration of key customers and suppliers	Business Process Transformation

Figure 5.2.   Approach to BPM in Supply Chain: Level 2 to Level 3

This chapter explores who should manage supply chain processes to gain higher value and how, once the firm has captured and deployed those most important internal process improvements for mutual success. The approach to be used is illustrated in Figure 5.2. As the firm truly moves from level 2 to level 3, the concept of sharing knowledge in a trusting atmosphere with a select group of business allies becomes the key ingredient to discovering all of the hidden values that have so far eluded the relationships, some of which have endured for scores of years.

Collaboration and technology are applied as the external cultural barriers are subdued and together businesses improve processes and infrastructure in parallel, as they seek the best possible cycle times for process execution, and the means to generate new revenues, and better utilize collective assets, ensuring that the best-able company performs the key process steps. Business process transformation occurs then in an atmosphere where key customers and suppliers are contributing directly to the process improvements that enhance performance across the extended enterprise.

BUSINESS PROCESS MANAGEMENT BECOMES THE KEY ENABLER AT THIS LEVEL

It is essential, as we describe how new values can be created and shared in level 3 and beyond efforts, that the reader appreciate that there must be an electronic means of transferring data and knowledge between the collaborating businesses. Until recently, such transfer was limited by the unwillingness of businesses to share data externally and because of the disparate systems in which the knowledge was stored, inhibiting easy access or transfer. BPM and business process management systems (BPMS) enter this scenario, as firms find the value of sharing previously secret information and analyzing the data and trends together to define new business propositions. Connecting enterprise-wide resource planning systems in a way that allows the participating firms to extract valuable information, without violating security issues, becomes the breakthrough technique for building a network response to market and customer needs.

Business agility develops as more than an interesting concept in this level, when BPM creates the opportunity to quickly and accurately share information between firms of any size, on a variety of subjects, and align these firms with business objectives. Document transfer moves from days to seconds, trade settlements are reduced from 5 days to an hour, build-to-order products are completed in 24 hours rather than 5 to 6 weeks, and call center inquiries are satisfied in 10 seconds instead of hours. These are just a few of the new capabilities brought on as the firm moves to the advanced levels of supply chain management and enhances its ability to communicate with selected business allies in an electronic environment, through the use of BPM technologies.

IMPORTANT ISSUES MUST BE ADDRESSED

To get started, the firm begins to decide with whom it wants to collaborate. This decision is approached slowly and carefully, as sharing information with a business ally, which in turn uses that knowledge for its own purposes or turns it over to a competitor, is the swiftest way to kill any external collaboration effort. We suggest the selection of one or two key suppliers and one or two key customers, where a pilot operation can be established to prove the validity of the concepts being tested. A business unit led by a visionary with a desire to become one of the first businesses in an industry to reach level 4 or beyond is a good candidate for such an effort. This leader should form a small guiding council, staffed with key internal members and a few volunteers from the supplier or customer. This council then plots the processes involved in the relationship and begins an effort to redesign the end-to-end system, bringing best possible practices to each important step.

This council will generally get off to a good start, discovering there are ways in which the most advanced organization can benefit from the learning of others, in areas often neglected in the typical supply chain improvement efforts. One firm, for example, may be particularly skilled at warehouse management, while another has strong abilities in planning and scheduling or order management. The sharing, on-line and in real time, of what can benefit both parties tends to bubble to the surface, as the council establishes teams to ferret out better techniques and practices. Sharing knowledge accurately and electronically becomes a central ingredient in this collaboration, and a new level of savings is generally found.

Among the questions and issues that will surface, which require attention by subgroups, are the following:

- Where are the components of information that can be shared, without violating security concerns, which can enhance the network's ability to provide better solutions to customers and enhance the processes designated for improvement?
- How can BPM tools be leveraged to enhance the strategic plan or the firm's business model (e.g., to develop new and better offerings into the market in shorter time cycles and with a greater probability of success)?
- How can BPMS be applied to retrieve components of knowledge that will help customize processes to achieve higher level benefits, in terms of flexibility, response time, and individual customer needs?
- How can an extended extranet, or "supernet," be established between the key business allies to transfer critical information through a visible pipeline of data transfer?
- How can this supernet be used to combine the possibilities introduced through B2B and B2C portals, public and private auctions, specific customer applications, order fulfillment service, and so forth?

In the new business environment, virtually every firm is engaged in some form of extended enterprise system, often in a multiplicity of such systems. Under these conditions, it is imperative that the fundamental business processes are handled as effectively as possible and that the linked business allies are working in a coordinated manner to deliver the highest level of satisfaction to the end customer or consumer. We see no way to accomplish that objective without a modern, state-of-the-art communications system. BPM has emerged as the tool set for delivering the most effective technology-enabled business processes. It essentially becomes an issue of finding the way to be a part of such collaborative network systems or losing ground to more enabled competitors.

Through a BPMS, a firm can find the means to operate its business in a more effective manner. For an explanation of the central ideas and basic elements, see *The Networked Supply Chain* (Poirier et al., 2004). The process steps across the extended enterprise can be defined and integrated in a seamless manner, allowing the participants to determine which entity is best able to perform the actual process. Then, using specific business rules, agreed to by the players, the processes can be optimized and monitored and managed by the appropriate BPMS. The traditional monolithic methodology, in which a single platform or software provider is used for all applications, is replaced by an explicit and layered model, which we will describe. Process flow is controlled in accordance with the accepted business rules, and better products and services are delivered to the customers.

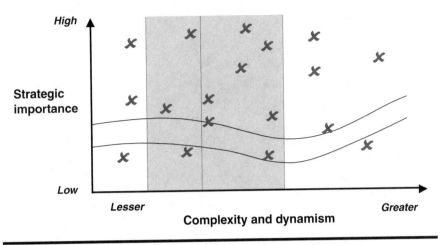

Figure 5.3. Understanding Processes

BUSINESS PROCESS MANAGEMENT SYSTEMS UNLOCK THE REAL BENEFITS FOR THE SUPPLY CHAIN

Supply chain processes (along with other business processes) can be analyzed in the context of the organization's strategic imperatives and evaluated from the perspective of both strategic importance and complexity and dynamism. Figure 5.3 shows a typical analysis:

- The strategic importance dimension relates to how crucial that process is in achieving the organization's competitive strategy. The higher the rating, the more influence the process has on achieving the strategic imperative.
- Complexity is a measure of the number of steps and/or participants in the process with reference to the number of organizational or departmental boundaries the process crosses. The greater the number, the more complex the process.
- Dynamism is a measure of the frequency of changes to the process that the organization would ideally want.
- The complexity and dynamism axis is an aggregation of these two factors.

The key strategic processes are those in which an enterprise would:

- Invest most management time and resources
- Reorganize around, to minimize handoffs and make the processes flow most efficiently

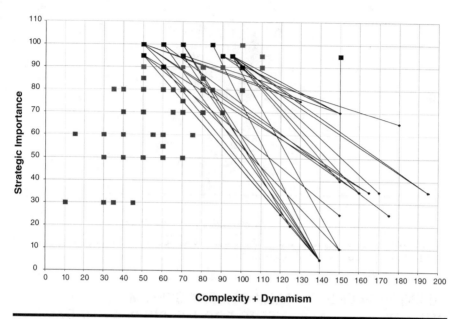

Figure 5.4. Mapping Current and Future State Processes

■ Retain as core to its success and which should remain in-house at all costs

The processes of least strategic importance are those that may be considered for alternative sourcing. The processes to the left of the chart are those that would typically be delivered through packaged solutions. Those in the middle of the chart would be delivered through work flow and/or rules engines. The most complex and dynamic processes, to the right of the chart, will, in the short term, largely remain manual.

An examination of the current and desired future states of the cross-enterprise processes reveals where the key opportunities lie and shows the prime targets for joint investment. Figure 5.4 shows a typical mapping. The current and future state mapping carried out under this approach also provides the organization with a number of other benefits:

■ Provides a more scientific and mathematical, rather than emotional or "intuitive," approach to change in any domain. Our experience shows that this is particularly true in the technology and organizational domains, where often many projects are spawned with indeterminate business value.

- Provides a way to drive the transformation plan — process by process — that maximizes business value and minimizes the delivery time for the organization's strategic imperatives.

The next stage is to link the current condition to the future state. In our experience, the mapping of existing processes to their future state is one of the most powerful drivers in determining the priorities for process transformation. The example in Figure 5.4, from a recent client project, shows a typical mapping. Many of the legacy processes prove to be overly complex and receive too little management attention compared to their strategic importance. The map immediately shows the priorities for process transformation:

- Investing in processes according to their strategic importance will ensure the greatest return in that investment.
- Eliminating unnecessary process complexity and dynamism will maximize people's productivity and reduce both the time to execute and the number of errors that occur.

BARRIERS — AND WHAT TO DO ABOUT THEM

In a survey, Forrester (2002) asked respondents what is dragging down supply chain performance. The highest response, 46% of interviewees, cited difficulties in getting processes and people adjusted to changes (see Figure 5.5). The survey quotes several companies:

- "Internally, we lack a common view of demand because our customer service group is separate from our supply chain organization" (construction company).
- "None of our apps — except our transportation system — can extend to partners, so we still rely on phone/fax/email" (healthcare company).
- "Implementing SAP's SCM apps involves tremendous business process changes and major employee training efforts" (chemicals company).

Clearly, the two key barriers to successful supply chain improvement are the ability to link processes through the organization and, beyond that, to link processes to suppliers and customers to enable real collaboration.

So what do we need to do to overcome the barriers and make the leap over the wall from level 2 to level 3? The authors believe that the activity falls into four broad areas:

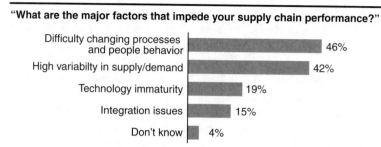

"What are the major factors that impede your supply chain performance?"

Difficulty changing processes and people behavior — 46%
High variabilty in supply/demand — 42%
Technology immaturity — 19%
Integration issues — 15%
Don't know — 4%

**Base: 26 supply chain executives at $1 billion-plus companies
(multiple responses accepted)**

Figure 5.5. Forrester Survey Results (Source: Forrester Research, Inc., December 2002)

- At the *business* level, there must be a willingness at the highest level to drive collaboration both within the company and across the company boundaries.
- At the *process* level, the company must be willing to develop processes in a conscious effort to align them with its strategy. A process strategy is needed, which will, in turn, drive many other strategies in the business — organization, applications, data, and technology.
- At the *organizational* level, the organization must be designed to make the most strategic processes as effective as possible. Thus, organizational boundaries will change so that strategic processes have as few handoffs as possible.
- *Applications, data, and technology* must also be aligned to support the strategic processes, within and across the organization.

MAKING THE TRANSITION TO LEVEL 3: THE BUSINESS LEVEL

Given the willingness of the senior management of the company to embark on the level 3 stage of the journey and a firm foundation of level 2 in place, the next step is to open the discussion with key suppliers and customers. We say key suppliers and customers because the discussions and effort needed to make real progress are not insignificant. Therefore, choose the partners that also have a willingness to explore the possibilities for finding greater value together, and limit the discussion to those where you believe the biggest value proposition lies.

The conversations with the chosen partners should include the examination of all aspects of the customer/supplier relationship, including technical, transactional, procurement, and logistics, along with product, information, and cash flows. Based on this type of analysis, begin looking for the hidden values across the full supply chain network that connects the customer and its supplier to the final consumers.

The real breakthrough comes when you can create win-win solutions that make more money for both firms. By defining specific opportunities for action, chartering improvement teams, defining success for them, and developing performance measures to track team performance, the value between the enterprises can be unlocked.

MAKING THE TRANSITION TO LEVEL 3: THE PROCESS LEVEL

By capturing and analyzing the inter-company processes, to identify those with the greatest strategic importance for both companies, the most profitable areas for investments can be isolated and prepared for improvement. It is clearly not possible to work on all processes; a degree of prioritization is vital. Typically, the processes we see being put at the top of the list are procurement based: auctions, catalogue-based inquiries, transaction automation, and, increasingly, checking schedules to ensure that an order can be fulfilled on time, in full.

More advanced tools for BPM, such as Intalio's tool set, will become an essential part of the level 3 company's tool kit. If processes on both sides of the inter-company gap are automated and made to comply with the same process protocols, then making them talk to each other will be relatively simple.

TRANSITION FROM LEVEL 2 TO LEVEL 3: ORGANIZATION

The organization of a process-based company will start to reflect the process priorities adopted. Departmental boundaries will move, and managers' roles and responsibilities will change, to minimize the disruption to the efficient and effective execution of the company's strategic processes. The authors are already seeing the impact of this in the early adopters.

For the level 3 company, the inter-company processes will be seen as having increasing importance. Organization will change to make the interfacing processes as effective as possible. There will be a few critical points of contact, well-defined data structures, effective problem resolution routines, and, above all, a clear view of the value added by making these arrangements.

As Gartner recently said: "Enterprises should begin to take advantage of explicitly defined processes. By 2005 at least 90% of large enterprises will have BPM in their enterprise nervous systems (0.9 probability). Enterprises that continue to hard-code all flow control, or insist on manual process steps and do not incorporate BPM's benefits, will lose out to competitors that adopt BPM" (Harris, 2004).

TRANSITION FROM LEVEL 2 TO LEVEL 3: APPLICATIONS, DATA, AND TECHNOLOGY

It is a curious condition that this trio is left to the last in the list. For many years, applications have been the starting point for investment and business improvement efforts. Latest trends, supported by our recent supply chain survey, show that this is no longer the case. Fewer companies are investing in applications. Rather, they are investing in making those they already have work properly and filling in the gaps — driven increasingly by the need to make processes work more effectively. So it is that the processes that drive the improvements and the role of applications become one of supporting the processes. In the authors' experience, up to 85% of any company's portfolio of IT projects are canceled or redirected once the correct process focus can be identified.

TRANSITION FROM LEVEL 2 TO LEVEL 3: THE BENEFIT CASE

With many companies striving to reach level 3, there must be a benefit case. As we have seen, some companies still do not believe in this principle. The authors firmly believe in, and have seen, the benefits of collaboration "across the enterprise boundary," from business unit to business unit or company to company. Technology has removed many of the barriers, for example, to true collaborative planning, forecasting, and replenishment (CPFR). The benefits come in many guises:

- Revenue improvement (+5 to 15%)
  - ☐ Resulting from reduced out-of-stocks and improved category management
- Cost reduction (15 to 30%)
  - ☐ For example, automotive manufacturers cut the cost per car by $5,000 through better collaboration between consumers, suppliers, and other partners (CSC automotive industry survey, 2001)

- Inventory reduction (25 to 50%)
  - ☐ Improved forecasting to replenishment between manufacturers and distributors, resulting in up to 25% lower inventory carrying costs
  - ☐ Dynamic resource allocation among contract manufacturers and logistics carriers, resulting in 30% lower work-in-progress inventories (Aberdeen Group, 2002)
- Improved customer service levels
  - ☐ Reduced cycle times (increased cycle turns), resulting from improved scheduling and planning
  - ☐ A 15% increase in fulfillment levels and customer satisfaction, resulting from committing to viable schedules based on reliable available-to-promise, capable-to-promise, and internal profitable-to-promise data and up-to-date resource capabilities (Aberdeen Group, 2002)
- Improved return on technology investment
  - ☐ Investments in traditional supply chain planning can be extended to external trading partners with minimal new investment
- Improved trading relationships
  - ☐ CPFR partners make commitment to measuring and sharing financial benefits

The results of the authors' survey of over 40 case studies for the transition from level 2 to level 3 are summarized in Figure 5.6. Again, significant gains were recorded though process improvements in the five SCOR® processes. While the overall percentage savings recorded were, in some cases, not as great as those for the level 1 to level 2 transition, when you consider that these new savings were built on top of savings already banked, the case looks impressive.

GENERAL MOTORS SHOWS THE WAY

As an illustration of how a firm can move successfully into level 3 and beyond, consider what happened at a venerable organization called General Motors, after chief information officer Ralph Szygenda was brought in to help GM regain some of its preeminence in engineering and car styling by "better applying and integrating information technology." As reported in *Supply Chain Management Review,* the first phase of the effort was launched in 1997, to improve the vehicle development process. There were two subparts to the initiative: reengineer design processes and implement time- and effort-saving IT. As progress was made, GM further realized that it really needed to embark on a supply-chain-

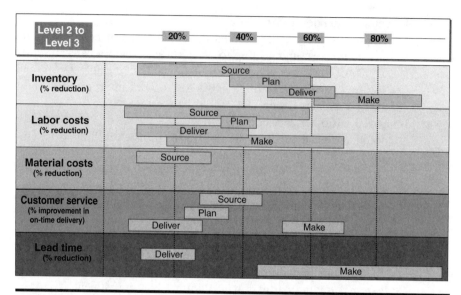

Figure 5.6. Moving Up the Supply Chain Maturity Model from Level 2 to Level 3

wide effort, extending much of the effort to suppliers willing to collaborate electronically. The results have been dramatic: a 70% reduction in product development cycle time and more than $1 billion in savings. As Szygenda reported, "Broadening the vehicle development initiative to include our external partners was always in the back of our minds" (Gutmann, 2003, p. 35).

When GM began its work on the vehicle development process, it formed teams to look at key design processes, to find how it could make them better and leaner. Concurrently, the teams looked at what IT could do to enable the improvements. This is a very typical step in the maturity process, as collaboration and technology are used to reach the higher levels of progress. The company started by asking tier one and further upstream suppliers to identify the problems with the existing design process. When this step is taken, most firms find there is a lot more that could be accomplished with existing processes than normally believed. GM used these conversations to identify three main areas of concern: batch processes that needed to be more interactive, a balance between security and accessibility, and improvements to the integration of suppliers and the transmission of data in a way that would not disrupt business processes.

Design files were synchronized with suppliers, ensuring that all parties were working from the same information. Instead of exchanging design tapes back and forth, an electronic solution was introduced to do direct computer-aided design (CAD) and computer-aided-manufacturing (CAM) exchanges. A common technology backbone was used to integrate disparate technology solutions and systems, "while providing one common source of up-to-date information related to a product" (Gutmann, 2003, p. 38). This move led to the adoption of one common CAD/CAM system at all GM locations and joint ventures across the world. A product data management system was also installed at that time.

In 1999, GM began connecting its suppliers to the common technology backbone through its virtual private network and its supplier portal, Supply Power. Through this technique, GM increased its suppliers' "access to business and technical systems such as project schedules, engineering change requests, technical specifications, test requirements, and quality controls. Tier one suppliers focused on keeping the design changes lean, passing on needed process improvements to their suppliers. Today, GM is able to integrate and update [its] bigger suppliers' CAD creations on a daily basis. For smaller suppliers, that information is updated weekly." Security and access to information are balanced by providing "different layers of visibility to different suppliers" (Gutmann, 2003, p. 36). At the end of 2002, GM had 1,000 suppliers in North America and 1,000 suppliers in Europe integrated into this new system.

In 2001, the scope was expanded to include external suppliers for such downstream processes as tooling, manufacturing engineering, and manufacturing. A tooling supplier can now download design files right after the design/styling phase, allowing it to begin its work earlier in the process. This assistance is offered through GM's product development portal, eliminating a previously heavily manually intensive set of processes. This feature allows GM to share facility and process design with its "most trusted suppliers," which in turn "accelerates the assembly design process" (Gutmann, 2003, p. 38). The result has been a reduction in product development process time from 60 months in 1996 to 18 months in 2003, a dramatic example of how collaboration will work, especially when enabled with the correct technology.

HARLEY-DAVIDSON'S EFFORT PAYS OFF

Some firms draw on knowledge and make it available electronically, as part of an effort to bring enterprise efficiencies to their business allies. Harley-Davidson took three years to develop and introduce its H-D Supplier Network. Companies making everything from valves and gears to pedals and clamps can enter this site and see production forecasts, account status, and two-dimensional drawings of parts. Suppliers can also submit shipment notices and receive inventory replenishment alerts, much faster and at less cost than previously. This portal eliminates the expense of an electronic data interchange system and reduces the time spent on such interactions from the typical 45 minutes to less than 15 minutes. "Nearly 300 of Harley's 695 parts suppliers now log on to applications through this supplier portal" (Sullivan, 2004, p. 47).

Known for its leading-edge practices, Harley has worked hard with its key suppliers to develop systems and methods that benefit both parties. Through the portal, which is provided to large and small firms, suppliers can manage payables and receivables so closely that a company can get paid within seven days of submitting an electronic invoice. On the other side, using a single material resource planning system, Harley provides its suppliers with the kind of information that leads to better handling of schedule changes, improved responsiveness to the unexpected, better planned capacity, and improved delivery performance. The parts centers, known inside Harley as Materials Velocity Centers, are a key component of this portal access. Requirements for the several manufacturing sites and the headquarters operation are coordinated so that a single reliable message is relayed to those needing to know what is actually happening, a key in any advanced-state supply chain communication network.

CONCLUSIONS

Moving effectively to advanced levels of supply chain progress is not an easy task, but the results are well worth the effort. A compelling amount of evidence is showing that the returns on effort can result in five to eight points of new profit for the participants. The cultural barriers and indifferent thinking that inhibit this progress must be overcome. This is accomplished best by establishing test or pilots with willing and forward-thinking business leader who want their business unit to be at the forefront of what advanced supply chain management can provide. As we go further, we will begin to elaborate on how BPM and BPMS become crucial ingredients for attaining the highest level of progress that makes sense for the business and its markets.

6

ACHIEVING DOMINANCE AT THE HIGHER LEVELS OF PROGRESS

We are now entering the rarified area of supply chain management, the most important part of the progression. When a business, usually a nucleus firm or channel master, completes its internal work and drives with its network partners into level 4, an area is entered where market dominance can be achieved. It is in this level that members of a supply chain network begin applying a variety of collaborative tools to create end-to-end visibility into the key threads of what has become a total value chain or networked enterprise. The conditions of this progress are won at a price. Investment decisions are made in conjunction with the desires and needs of value chain partners, and the processes — especially those of a financial nature — must reflect the appropriate conditions of sensible investment with adequate return on investments. Process flexibility is a key in this level of enterprise interaction across multiple networks.

RESULTS PROPEL A FIRM AND ITS BUSINESS ALLIES TO HIGHER PERFORMANCE

In what becomes a *co-managed value chain,* the overarching focus is brought to value propositions for the business customer and end consumers that cannot be matched by competitors. Seamless electronic linkages are pervasive, carrying the kind of data and knowledge that cannot be garnered from any other business network. Business process management (BPM) initiatives are present here, as they become a high-priority item for those companies that desire to use BPM to share vital business information, particularly from their enterprise resource planning (ERP) systems, and find the next level of improvement. The dual purposes are to increase operational performance and maximize the return on the total assets employed in the extended enterprise. Businesses that have learned the value of trust between a few carefully selected business allies use BPM tools here, to evaluate and optimize their joint business processes and to eventually create and nurture their competitive advantage against slower and less able networks.

An example of the kind of progress that can be made is given by beer and alcohol manufacturer Diageo plc. In a move aimed at increasing revenues, this firm determined that North American network partners should not run out of its popular Guinness beer. The firm decided to go through a supply chain improvement effort that would include sharing of "real-time sales and replenishment data with distributors." The motive was to encourage these distributors to work harder at sales and less at generating redundant analyses and worrying over missing inventories. With the help of Manugistics Group, Inc. and that company's collaborative planning, forecasting, and replenishment software, the firm linked its distributors into manufacturing sites in Ireland, the Caribbean, and Africa. The system helps Diageo "combine weekly data from sales, on-hand inventory, and distributors' receipts with promotional information to generate detailed forecasts that are automatically sent via a Web application to 120 distributors, which represent 80% of Guinness' business." (Bacheldor, 2003, p. 57).

With this information delivered on a timely basis, a real form of collaboration can take place. Differences resulting from interpretation of the forecasts can be resolved electronically, and when agreement is reached, the accepted numbers can be loaded into a replenishment system that creates inventory levels and sales. Diageo "expects to save up to $1.1 million in inventory reduction, and reap some $600,000 in logistics benefits, with sales being boosted by 1%" (Bacheldor, 2003, p. 57).

The ultimate objective of BPM and business process management systems (BPMS) in this level becomes the capability to connect business firms and important functions within those firms in order to gain a distinctive advantage in the eyes of the most important customers and consumers. From sourcing to customer care, BPM is used by multiple organizations to transfer crucial business knowledge and enhance those processes of greatest importance first to the supply chain constituents involved and then to the intended customer base. Better coordination, shorter cycle times, errorless transactions, and optimized process conditions become the metrics of importance in this level of progress. Ultimately, the business allies find the means to increase operational performance and further develop their competitive advantage.

ADVANCED SUPPLY CHAIN MANAGEMENT REQUIRES A COORDINATED PLAN AND ENABLING TECHNOLOGY

As a firm and its most trusted and closest business allies prepare to enter this fourth level of the maturity model, they realize that a solid plan of attack is necessary for success. It is not enough just to link firms that have completed their internal homework and begun to work with external partners. There must be a coordinated effort aimed at optimized conditions across the extended enterprise and a willingness to share the vital knowledge that leads to differentiating processes. Advanced supply chain management, therefore, cannot be accomplished without a clear roadmap that identifies where the network is headed and how the participants can measure the progress made and show the results on the financial statements.

A series of obstacles will be encountered as this planning begins. First, there will be the natural resistance to sharing valuable data over any electronic system. This problem is dealt with by an early agreement on what can and will be shared, what will be kept protected and proprietary within the various businesses systems, and what security features will protect the data that are not to be shared at all. Second, some of the participants will view the sharing of knowledge as valuable, but will balk at the work necessary to bring BPMS to their existing methodology as they lack the understanding of the values or perceive the application as too difficult to accomplish.

Since all of the key players must be involved in network connectivity, time should be spent to train, if necessary, and to develop the capability of all constituents to participate in the data sharing. Key to this is the identification and, if necessary, simplification of the interfacing processes. We strongly advise the establishment of diagnostic labs or demonstration labs to discuss, explain, and show the potential values to be added from such data sharing. These labs serve the dual purpose of explaining what activities the participants will be pursuing, while allowing them to test their hypotheses and get a feel for the end results. In Chapter 9, we will explain how the participants can go further and simulate the changed processes that result from the knowledge sharing, before investing too heavily in execution.

Third, initially there will be poor understanding of how to deploy the tools and methods, resulting in poor definition of what the early gains might be and delays in execution. Once again, time spent on clarifying the roadmap and establishing a set of driving metrics will aid immensely in getting the necessary buy-in and provision of key resources to move forward. Fourth, there will be some confusion regarding the central purposes, especially as some participants view BPM as a means to drive automation and eliminate personnel, rather than the route to process optimization and the best possible customer satisfaction.

Each of these complications can be defeated through an adequate plan for BPM deployment and a roadmap that defines the resolution of fundamental business issues plaguing each of the network members. While most of the solutions will derive from interactive discussions and planning between the participants, as will be described shortly, several software suppliers can add value in this part of the effort and could play a role in development of the plan. As mentioned, Manugistics was very helpful to Diageo.

Fuego, Inc. offers another example. This firm provides "templates for more than 60 sets of processes across the financial-services, insurance, manufacturing, and telecommunications industries." Pegasystems, Inc. has "rolled out a series of process templates that accelerate deployment by automating common back-office processes while letting companies manage exceptions" (Greenfield et al., 2004, p. 68).

For our purposes, we have selected Yantra as a service provider to follow, as we use its experiences and cases studies to document what can be done across enterprises and to establish the kinds of improvements that can be deployed and the type of savings to be found after the effort is completed. Yantra has headquarters in Tewksbury, Massachusetts and has a proven track record with such firms as Best Buy, Allders Department Store, Rockport, Target, and US Transcom. Its extensive suite of applications runs the gamut of supply chain needs, from distributed order management and flow through provisioning to synchronized fulfillment and reverse logistics. As we develop our deeper analysis into the final levels of the maturity model and the best use of the SCOR® model, we will use Yantra as an example to bring a sense of reality to what can and does happen.

PROVIDING ADDED VALUE FOR CELLSTAR

We begin with a case involving Yantra, as a provider of software for systems integration, and CellStar, a provider of value-adding logistics and distribution services for the wireless communication industry, headquartered in Dallas, Texas. CellStar was once a part of Motorola, responsible for the packaging and delivery of handheld telephones to a wide variety of industrial and consumer markets. At the time of this case, the firm was independent and competing for business in the U.S. wireless market. This market was typified by a leveling of the handset growth curve and had become an industry increasingly viewed as offering a commoditized set of products.

CellStar, as a major provider of packaging and distribution services for such products, found itself in a situation in 2002 where it was competing to be a high-value strategic partner in a business environment characterized as a race for competitive advantage. CellStar's financial position was not robust, and the firm was finding it very difficult to generate cash in this business environment, where it was competing with other distributors such as Ingram Micro and Avnet; contract manufacturers like Flextronics and Solectron; with transportation service providers UPS, FedEx, and USPS; and third-party logistics providers Menlo and Innotrac.

The company had an ERP system, but it had an inherent problem that limited CellStar in its market ambitions, which included expansion of its customer base. The problem revolved around its ERP system having been designed as a "system of record" for a single enterprise process. As such, the firm was unable to compete for business where real-time order, inventory, and shipment visibility across the logistics service network was a requirement. XML messaging could be used to extend its ERP reach, but only for one-to-one processes between customer and provider. This condition created a reconciliation nightmare for inventory tracking and offered no support for processes that extended beyond arm's reach. As a result, the company had a high cost structure to support any many-to-many processes. Coordination of orders for product, service, installation, provisioning, and delivery was a serious challenge. The firm was simply not able to meet core requirements of new opportunities or respond adequately to new business opportunities beyond wireless applications.

The business situation, which brought these conditions to a point needing resolution and led to the resultant improvement, occurred when CellStar faced an opportunity to provide Alltel, Cingular, and Sprint PCS with outsourced call center, fulfillment and returns services. The initial offering was to cover upgrade and replacement fulfillment of all equipment requests. Typical of today's complex delivery environment, the basic requirement was to support a virtual and extended enterprise business model, from carrier stores and individual customers through call centers, fulfillment sites, repair sites, and final product disposition. The chosen provider, CellStar was told, would need to deal with complex process rules, based on customer preferences and product availability in multiple inventories. Time to market was a key metric that had to be reduced or the company would lose the opportunity.

It became a joint sales opportunity for CellStar and its selected business ally, Yantra. CellStar was to provide the seamless end-to-end delivery system with the necessary process improvements, including fail-safe electronic linkage. Yantra would introduce mechanisms that would improve the capability of the ERP system and link it with suppliers and customers. The software and systems company would also provide a business-rules-driven fulfillment and returns engine, with rapid integration into CellStar's business processes and systems. Together, they would introduce the necessary visibility into the key threads in the new system, as well as provide event monitoring and early notification of any systemic problems.

Through the partnering that was to occur, CellStar believed it could provide an unmatched value proposition for the potential business customers and eventual consumers. The potential increase to CellStar revenue was estimated to be $33 million by 2004. The business allies began by drawing process maps of the end-to-end system that was to be enhanced and preparing a business case foundation for the eventual actions. Figure 6.1 describes this foundation, which used the project summary to establish the need for a program, dubbed Gatekeeper, which would provide the required systems and ERP extensions. The new processes would be aimed at wireless carriers and meet their need for complex logistics services, while introducing a "converting pipeline" that included capability to handle all fulfillment services and reverse logistics (a key element in the offering). Among the operational impacts, the system would control on-hand inventory and reduce selling, general, and administrative (SG&A) costs.

The firms viewed the work as process improvement that would progress through a multiphase deployment, as described in Figure 6.2. Business capa-

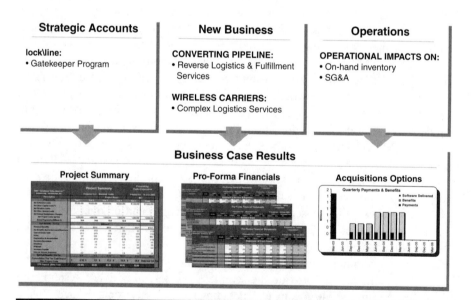

Figure 6.1. Business Case Foundation

	Today	Phase I	Phase II	Future Phases
Business Capabilities	Basic Distribution Limited Repair	CellStar Returns	Complex Fulfillment	Supply Chain Optimization and Complex Collaboration
Market Opportunities	Carriers Wireless OEMs Consumer Direct Agents/Master Agents	Carriers Technology Retailers OEMS & Distributors Insurance Replacement	Cable DSL Satellite Wi-Fi	Not Yet Defined Nor in Scope

Figure 6.2. Deployment Assumptions: Work Should Progress Through Multiphase Deployment

bilities would be matched with market opportunities in a timed manner, starting with the immediate needs and progressing to an advantaged future state. In a complementary fashion, Figure 6.3 shows value assessment steps that were included to make certain the enhanced processing would meet the predetermined objectives. Note the targeted capabilities demonstration in the middle of this analytical progression.

Particular attention was given to the proposed reverse logistics pipeline, described by one of the process maps, depicted in Figure 6.4. Bringing special capabilities in this important process step helped distinguish the final offering in the eyes of the wireless customers. Through the Gatekeeper program, Yantra provided a total call center and equipment fulfillment solution, with the ability to link into suppliers and customers for necessary information. One unexpected benefit of this solution was that the refurbished units in the replacements mix increased from 20% to 38%, which helped to meet the large demand backlog. The overall intention was to create a distinctive advantage for CellStar, using technology as a strategic asset, which was accomplished and helped the firm expand its sales effort into other business areas.

The partnering that took place has been a success, as the new capabilities met or exceeded the intended objectives. CellStar introduced its new offering under the banner name "Omnigistics." It was presented as a set of solutions that provided depth and breadth to a customer's logistics strategy, turning the supply chain into a competitive advantage. The suite included forward logistics, with procurement and inventory management, order processing and reporting, fulfill-

	Opportunity Analysis	Business Strategy Analysis	Solution Definition & Evaluation	Investment Value Analysis	Acquisition Analysis
Objectives	Benchmarking vs. industry to discuss areas of opportunity Capture initiatives & issues to increase competitiveness	Document detailed business issues behind initiatives & high-level issues	Evaluate key Yantra capabilities to enable resolution of priority business issues	Build realistic business case based upon your: Financials Issues Assumptions	Funding & implementation strategy
Deliverables	Benchmarking study & documented business strategy, issues & initiatives	Prioritized & documented set of business issues Preliminary financial & operational impact	Targeted capabilities demonstration	Exec/Board-level business case Business release implementation strategy	Terms-based proposal designed to fund highest impact business release strategy

Figure 6.3. Value Assessment Steps

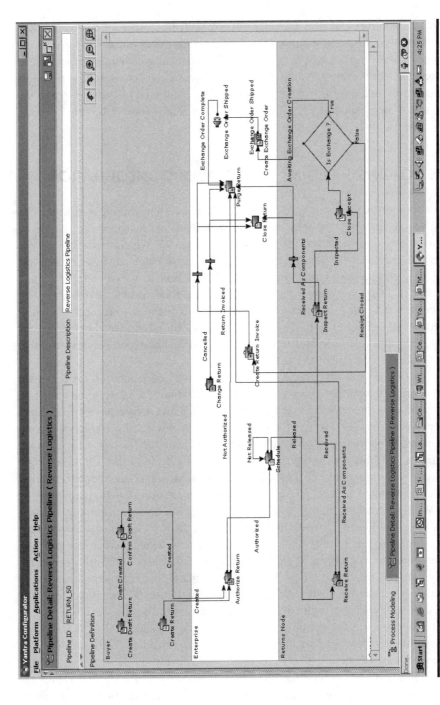

Figure 6.4. A Process System for Leveraging and Extending CellStar Capabilities

ment, and delivery. It also included reverse logistics, with returns, repair and refurbishment, disposition, and redistribution. Figure 6.5 illustrates some of the results of the effort. Stratascope, another business ally of Yantra, was used to prepare an "Investment Value Analyzer." The figure shows that the project summary included a five-year projection of the total cost of ownership and the revenue to be generated. Combined with lower SG&A costs, the impact to net cash benefits was very impressive.

THE ROADMAP BEGINS WITH PLAYER CONSOLIDATION

Collaboration as a business technique, particularly one to help a firm and its business allies progress into advanced supply chain management, has been widely discussed, with very mixed reviews, not all of which match the advantages described in the preceding case. While we acknowledge the difficulties in getting companies to truly cooperate with one another, we have found a small number of firms that have been able to achieve good results by working with a very limited number of very trusted business associates to reach higher performance. Michael Bauer, Director of CSC Lean Enterprise, has been very helpful in this research, and we will use much of his expertise as we outline how leaders accomplish the task.

The first step in that process is to discover why collaboration breaks down and to prepare a methodology that will work. We find that most of the problems occur because processes are often duplicated across an enterprise system, and the players refuse to leverage the best techniques, tools, and systems, preferring instead to support their own system, even when it is less effective for network purposes. At other times, the traditional linkages rely too much on technology, especially commercial off-the-shelf software and not co-developed enabling systems, oriented around BPMS and the particular needs of the network.

With an understanding of the need to conquer these essentially cultural inhibitions and to use collaboration effectively, the next move is to consolidate the critical processes, regardless of the number, into a single enterprise process, leveraging the best player at each step. Figure 6.6 is a pictorial illustration for achieving this condition, as the key players bring their expertise to bear on the overall system rather than steadfastly advocating their own solutions and systems. Across whatever represents the extended enterprise, the search is for the best processes, regardless of which firm is the originator.

As this search progresses, the players realize there must be a passage through five levels of cooperation, similar to the five levels of the maturity model. Through Figure 6.7, which was presented previously as Figure 5.1, Bauer explains what is being considered. In the early levels of the relationship, the firms

Project Summary

TVA™ Investment Value Analyzer™
Powered By : Stratascope, Inc.

Prepared by:
Yantra Corporation

Prepared For: USA CellStar   Project Name :   Printed On: 06-Mar-2003

Description	Year 1	Year 2	Year 3	Year 4	Year 5	Totals
Project Costs						
Net Software Costs	$1,080,000	$0	$0	$0	$0	$1,080,000
Net Other Capital Costs %	-	-	-	-	-	-
Net Services Costs	1,900,000	-	-	-	-	1,900,000
Net Other Annual Costs	-	-	-	-	-	-
Net Annual Maintenance Charges	237,600	237,600	237,600	237,600	237,600	1,188,000
Net Project Costs by Year	3,217,600	237,600	237,600	237,600	237,600	4,168,000
Annual Payments (Millions $)	**$3.2**	**$0.2**	**$0.2**	**$0.2**	**$0.2**	**$4.2**
Cash Benefits - Pre-Tax			Millions $			
Revenue Benefits	$5.8	$31.8	$37.0	$37.0	$37.0	$148.6
Net Benefits due to Increased Revenues	0.0	0.5	1.8	2.1	2.3	6.7
Cost of Goods Sold	0.2	7.9	12.2	12.2	12.2	44.7
SG&A	0.3	(4.4)	(7.4)	(7.4)	(7.4)	(26.3)
Depreciation & Amortization	-	-	-	-	-	-
Accounts Receivable	(0.4)	(0.5)	0.5	(0.0)	(0.0)	(0.5)
Inventory	2.4	0.4	0.6	0.4	0.4	4.2
PP&E Net	-	-	-	-	-	-
Accounts Payable	-	-	-	-	-	-
Interest Income (Expense)	0.1	0.3	0.4	0.4	0.5	1.7
Net Cash Benefits - Pre-Tax	**$2.6**	**$4.2**	**$8.2**	**$7.7**	**$8.0**	**$30.6**
Cummulative Pre-Tax Cash Impact After Project Costs	$ (0.7)	$ 3.3	$ 11.3	$ 18.7	$ 26.4	Historical OA

Total Cost of Ownership → (Net Project Costs by Year)

Revenue Generated → (Revenue Benefits)

Net Cash Benefits → (Net Cash Benefits - Pre-Tax)

Figure 6.5. The Results: A New Asset to Help Generate Cash and Enable Higher Margin Business

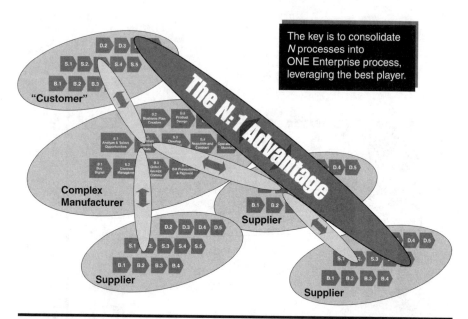

The key is to consolidate
N processes into
ONE Enterprise process,
leveraging the best player.

Figure 6.6. The N:1 Advantage©: Consolidation Rather Than Linkage (©Supply Chain Innovations, LLC)

work on *informing* and begin a sharing process that includes working together on design changes, electronic data interchanges, and electronic transfer of funds. As the *interacting* increases, they begin exchanging data on such things as new design introductions and how substitute materials can be used. In the *transact* level, the allies are conducting business in a fundamental but collaborative manner, including order entry and fulfillment, expediting to meet demands, and reporting on exceptions and returns.

In the fourth level, which is now being considered, the partners work on *delivering,* to the targeted customers and consumers, making the key players a part of the overall business effort. Now the provision of services includes joint diagnostics, review of reports, collaborative design and engineering, and joint inventory management. Visibility into the end-to-end systems becomes an essential element in this latter part of the effort. In the highest level, a *community* effort has appeared, and the players fully utilize their co-developed business and operating model to cross-company interactions and to personalize the solutions being brought to the targeted customers/consumers.

Figure 6.8 presents a small list of the companies that have successfully made their way through the progression, with the help of trusted allies, and the results that were achieved. Although these cases are mostly involved with early types

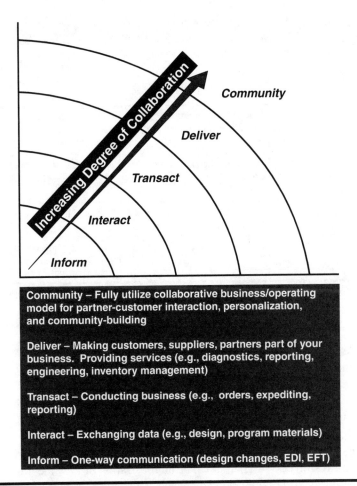

Figure 6.7. Understanding Levels of Collaboration (Source: Michael Bauer)

of collaborative effort (the design and introduction of new products and services), the magnitude of the improvements stands to illustrate what can be accomplished when firms bring their best and brightest together to enhance what they do together.

FIVE STEPS OUTLINE THE COLLABORATIVE PROCESS

With a determination to reap the benefits of collaboration and enhancement of the processing through technology, Bauer recommends a five-step process:

Company	What They Did	Results
Automotive Supplier	Collaborative product design	Reduced product development time by 50% — from 44 to 22 months
High-Tech Manufacturer	Linked product design and manufacturing across the Pacific	Reduced product development time and costs by 45%
Medical Equipment Manufacturer	Increased visibility of product data	Reduced time to produce kits by 85% — from 75 to 10 days
Manufacturer	Installed a PDM system and gave access to suppliers	Reduced build hours by 30%
Aerospace Company	Linked suppliers to designers	Mobilized a new project in 75% less time

Figure 6.8. Manufacturing Companies That Made an Impact Through Collaboration

- **Agree on the principles supporting the collaboration** — Without an understanding that the purpose is to seek higher level benefits for all participants, one or more parties are going to withdraw at some point in the effort to find the best enterprise processing. A simple single-page listing of the principles and the metrics that will be improved generally suffices to satisfy this requirement.

- **Establish a governance model to oversee the expenditure of time and resources** — There must be an illustration of how the mutual effort is going to be conducted, managed, and implemented.

- **Validate the internal processes** — As we have stated, progress to maturity starts inside the four walls, and when the house is in order, it can expand externally.

- **Consolidate processes** — Across a platform that moves from the individual company intranets to the network extranet, there needs to be guidance on how the efforts of the network will be consolidated for best results and how to transform the enterprise to an industry-dominant position.

- **Transform the enterprise** — With the purposes clear, the roadmap in hand, and an understanding of what is in the effort for all the key players, the stage is set for a major transformation effort that establishes the network as the one to beat in any industry.

Step 1: Agree on Principles

As the collaborative processing begins, the parties come to a table and agree on the principles that will guide their joint efforts. Figure 6.9 is an illustration Bauer uses to help this effort. The parties to this collaboration took the time to list the enterprise process steps, much as the SCOR® model would do, and then to overlay the principles. Note that it begins with sharing the benefits and rewards, an essential to success. The balance of the ten principles is self-explanatory, but forms the backbone of how the effort will proceed. Yomi Famurewa, vice president of supply chain for ArvinMeritor™, a leading tier one automotive supplier, and his team used these principles to make significant improvements in product development cycle time. Famurewa and Bauer shared these insights with an enthusiastic audience at Supply Chain World in March 2004.

Step 2: Establish a Governance Model

With the principles established, the players can move to setting up a governance system that will assure best results. Figure 6.10 describes such a model. Oversight is given to a collaboration council, but with a supplier advisory board nearby to give counsel on how to take best advantage of what this group is willing to contribute. The corporate executive committee of the nucleus firm or driving entity behind the effort must also play a part, to make certain the business plan objectives are being met. Design teams, assembled from the various business units and willing business allies, are then established to perform the actual collaboration and move the effort to specific objectives. Notice that this council determines priorities, licenses projects, sets policy, polices the players, and manages the enterprise bonus pool or how the benefits will be dispersed. This list of control features needs to be analyzed and made specific to the collaboration being conducted.

Step 3: Validate the Internal Processes

The next step is designed to assure the effort is starting off in the right manner. Figure 6.11 is intended as a reminder that the internal processes need to be the best across the total internal organization and meet the demands of being best in class from an industry perspective. This thinking is fundamental — you cannot expect your external allies to share equally with you and reach optimized conditions if your own processes are broken. There must be a reasonable level of parity with regard to key processes and the ability to add value for all constituents before embarking on a serious collaborative effort. Moreover, there

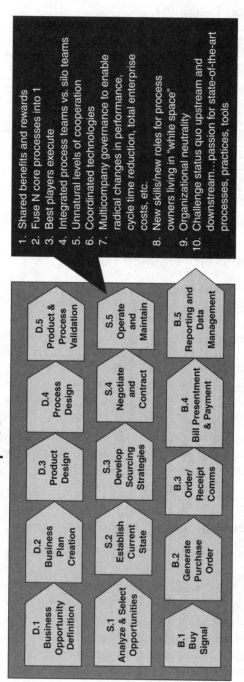

The Enterprise
D.1 Business Opportunity Definition
D.2 Business Plan Creation
D.3 Product Design
D.4 Process Design
D.5 Product & Process Validation
S.1 Analyze & Select Opportunities
S.2 Establish Current State
S.3 Develop Sourcing Strategies
S.4 Negotiate and Contract
S.5 Operate and Maintain
B.1 Buy Signal
B.2 Generate Purchase Order
B.3 Order/ Receipt Comms
B.4 Bill Presentment & Payment
B.5 Reporting and Data Management
1. Shared benefits and rewards
2. Fuse N core processes into 1
3. Best players execute
4. Integrated process teams vs. silo teams
5. Unnatural levels of cooperation
6. Coordinated technologies
7. Multicompany governance to enable radical changes in performance, cycle time reduction, total enterprise costs, etc.
8. New skills/new roles for process owners living in "white space"
9. Organizational neutrality
10. Challenge status quo upstream and downstream...passion for state-of-the-art processes, practices, tools

Figure 6.9. Step 1: Agree on Principles of Collaboration

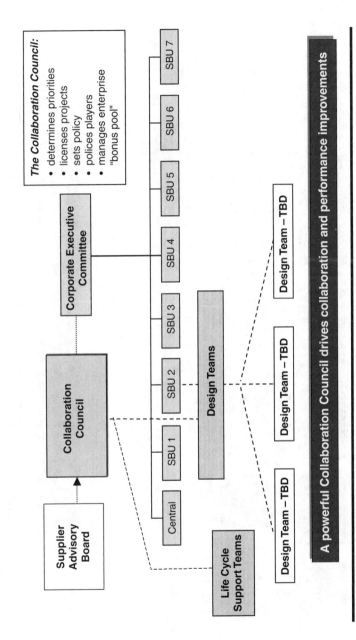

Figure 6.10. Step 2: Establish a Governance Model

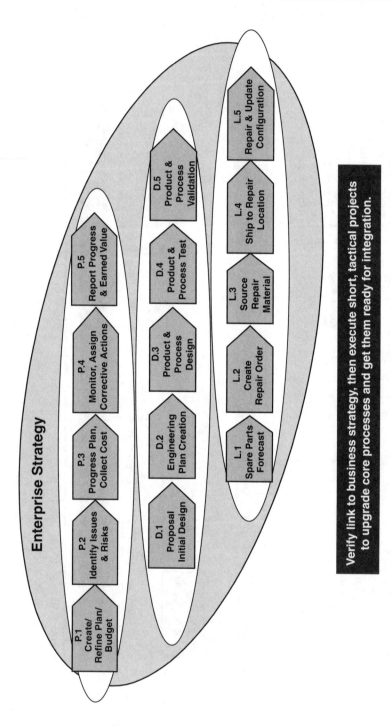

Figure 6.11. Step 3: Validate the Internal Processes — See That Key Internal "Silo" Processes Are Best in Class (*Get Your Own House in Order*)

must be a clear link between the improvement effort to the key processes and the business strategy.

Step 4: Consolidate Processes and
Step 5: Transform the Enterprise

The final two steps become the crux of the effort, but cannot proceed without the proper preparation. With the principles and governing structure in place, and the internal processes ready for external collaboration, the nucleus firm organizes the workshops, diagnostic labs, or team exercises that lead to consolidating the processing into the one best system. As described in Figure 6.12, the participants throw themselves into a serious and frank analysis of the as-is conditions and begin design and development of the improved could-be environment. When the new state has been created and accepted, then the crucial final step of transforming the enterprise takes place and, usually under pilot conditions, the new model is tested and amended if necessary.

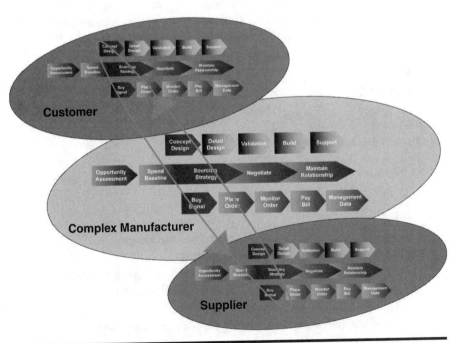

Figure 6.12.   Step 4: Consolidate Processes (N:1) and Execute a Pilot Project (*Make It Real*), and Step 5: Transform the Enterprise

ARVINMERITOR™ ILLUSTRATES THE POSSIBILITIES

ArvinMeritor™ is a Michigan-based U.S. manufacturer and distributor of automotive and truck parts and components. Bauer worked with this firm as it followed the five-step collaborative process, to achieve notable results, working with some of its key suppliers, with a focus on the automotive market. The corporation had a defined product development process and supply chain management process. It was also migrating from legacy ERP systems to a new system. Three business units and several joint ventures were involved, with 150 plants and 30 engineering facilities. The product lines included light vehicle, commercial vehicle, and aftermarket products. Among the issues faced were several design and supply chain systems, incompatible change management systems, a variety of bill of material techniques, and a rapidly proliferating parts list. The goal established for the change team was to create a collaborative platform that enables the stakeholders to plan, design, source, make, and deliver the products anywhere. The project approach was intended to establish the collaborative platform and implement a common product cost and change request system using the five-step approach.

As shown in Figure 6.13, the firm began by agreeing on the principles of collaboration. In this tripart model, suppliers were linked into a supplier management system, which further linked into operations and its ERP system.

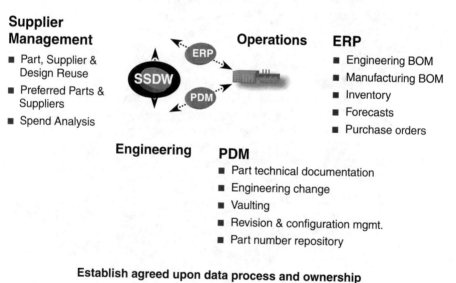

Supplier Management
- Part, Supplier & Design Reuse
- Preferred Parts & Suppliers
- Spend Analysis

SSDW

Operations   ERP

ERP
- Engineering BOM
- Manufacturing BOM
- Inventory
- Forecasts
- Purchase orders

Engineering   **PDM**

PDM
- Part technical documentation
- Engineering change
- Vaulting
- Revision & configuration mgmt.
- Part number repository

Establish agreed upon data process and ownership

Figure 6.13.   Step 1: Agree on Principles of Collaboration (Source: ArvinMeritor™)

Product data management (PDM) databases were used to link engineering as well, and the loop was completed through the strategic sourcing data warehouse (SSDW). The illustrated topics were considered areas where knowledge could be exchanged, and the group agreed on how the data processes worked, who had ownership, and how the data would be shared.

The governance model is illustrated in Figure 6.14 and shows that an engineering council was set up to oversee the effort, with linkage to the corporate engineering committee and a supplier advisory group. The engineering directors, in this case, were given responsibility for driving the effort toward a common process and setting the performance improvement goals. As shown in Figure 6.15, to validate the internal processes, they created a collaborative platform and moved from the key processes to common processes and systems, across the SCOR® model. All current processes were reviewed and critiqued so agreement could be reached on practicality, opportunity for automation, piloting, and then readiness for implementation.

Figure 6.16 depicts how the group then began to consolidate and transform the enterprise. The five blocks on top of the collaborative platform were deemed to be the keys to success. Beneath each heading are the elements involved in the redesign and transformation process. The goal was to develop an integrated intra- and inter-enterprise platform, which would be accepted as a continuous improvement process and not just an event. Figures 6.17 through 6.22 are illustrative of how each block was reviewed, redesigned, and transformed. Access to the details contained in these files was given to appropriate individuals, and all changes were handled electronically.

APPROACH TO BUSINESS PROCESS MANAGEMENT IN SUPPLY CHAIN: LEVEL 3 TO LEVEL 4

The transition from level 3 to level 4 requires a significant shift in thinking. Now companies must take account not only of what their immediate customers and suppliers are thinking, but also the thinking of others in their value chain, from ultimate supplier to ultimate customer. That is a pretty wide span, and in practice, a value chain runs back until the items can be considered as commodities. The level 4 investments in schedule synchronization and co-planning are unlikely to be made with companies that provide a true commodity product or service. In these cases, the general attitude is "If you don't like it, you can get it elsewhere!" This puts some bounds around the value chain processing.

So what is the impact? If we look at the six domains of change, we can summarize.

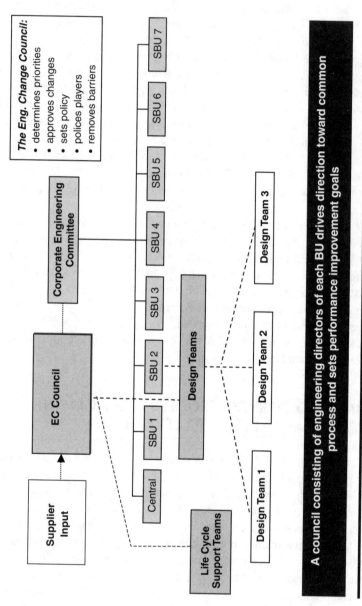

The Eng. Change Council:
- determines priorities
- approves changes
- sets policy
- polices players
- removes barriers

Corporate Engineering Committee

EC Council

Supplier Input

Central | SBU 1 | SBU 2 | SBU 3 | SBU 4 | SBU 5 | SBU 6 | SBU 7

Design Teams

Life Cycle Support Teams

Design Team 1

Design Team 2

Design Team 3

A council consisting of engineering directors of each BU drives direction toward common process and sets performance improvement goals

Figure 6.14. Step 2: The Governance Model (Source: ArvinMeritor™)

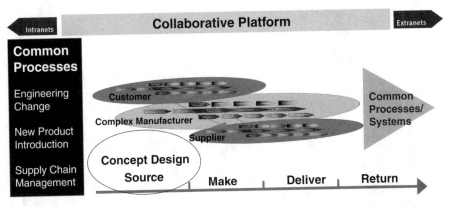

Review all current processes, agree on a common process if practical, automate, pilot, then implement

Figure 6.15.   Step 3: Validate Internal Processes (Source: ArvinMeritor™)

Processes

Level 3 companies will understand well the need to manage their process portfolio. The portfolio will identify clearly the processes that flow across boundaries. As they move to level 4, this understanding will be shared across their value chain partners. The interaction processes will be simplified; at the lower levels of maturity, there are often many touch points between companies, with the complexity adding cost and increasing the opportunity for error.

Business processes at level 4 extend across organizational boundaries. End-to-end processes at level 4 will flow across several boundaries. With the advent of automated process management systems such as BPMS, companies in the extended enterprise will increasingly automate their process interactions. The need for a process protocol is vital if flexibility is to be maintained. The protocols developed by organizations such as BPMI.org mean that processes can be described in a way that deals automatically with issues arising from the interfaces between enterprises' systems. From the point of view of the CEO of each company, this is largely a technical matter of little significance, other than to ensure the interface is standardized so that integration costs are minimized.

To be clear, private process instances, shared by businesses, use public processes to coordinate the interactions among them necessary to complete the desired overall business goals. In the BPMI brochure, the following words appear:

> BPMI.org considers an e-Business process conducted among two business partners as made of three parts: a Public Interface and two

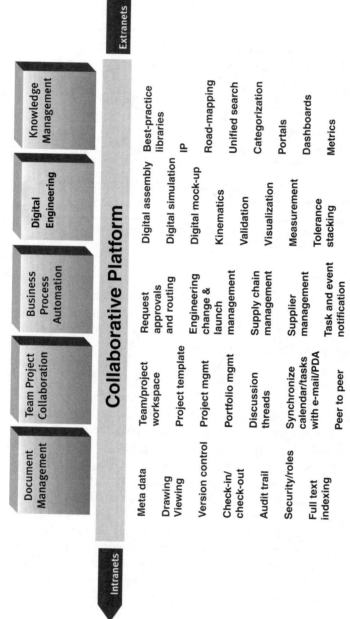

Figure 6.16. Steps 4 and 5: Consolidate and Transform the Enterprise (Source: ArvinMeritor™)

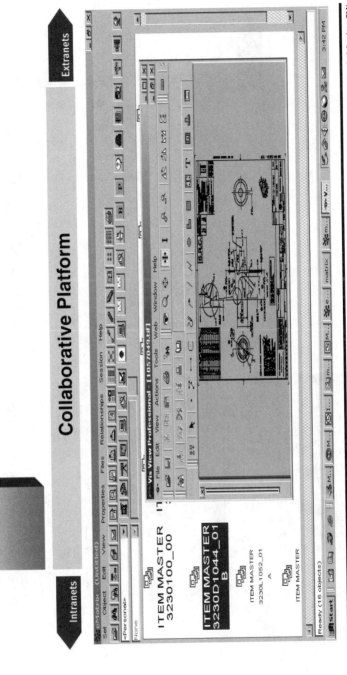

Figure 6.17.   Steps 4 and 5: Consolidate and Transform the Enterprise — Document Management (Source: ArvinMeritor™)

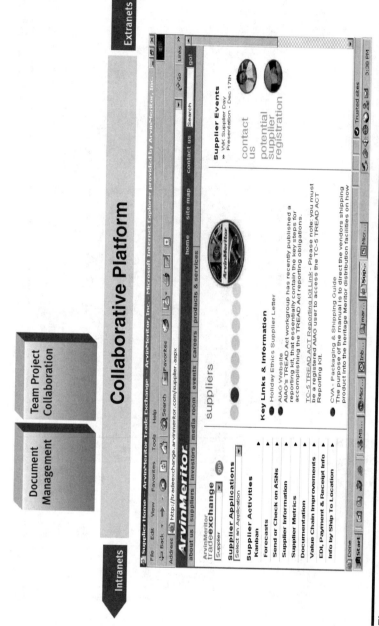

Figure 6.18. Steps 4 and 5: Consolidate and Transform the Enterprise — Team Project Collaboration (Source: ArvinMeritor™)

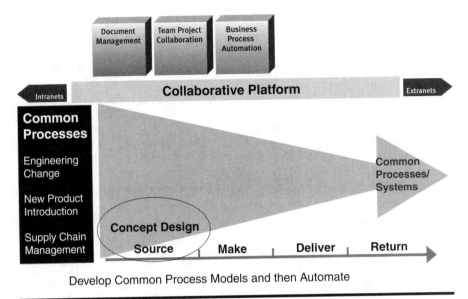

Figure 6.19.   Steps 4 and 5: Consolidate and Transform the Enterprise — Business Process Automation (Source: ArvinMeritor™)

Private Implementations (one for each partner). The Public Interface is common to the partners and is supported by protocols such as ebXML, RosettaNet, and BizTalk. The Private Implementations are specific to every partner and are described in any executable language. BPML is one such language.

Without an agreed and shared way of describing processes, changing from one partner to another in the chain or improving the operation of processes across the process boundaries becomes more and more complex and will be a real drag on the value chain's ability to improve and compete. With a shared protocol, the inhibitions are removed and the extended enterprise processes run straight through as though they are part of a single enterprise.

Best practice processes can be replicated across the value chain. Figure 6.23 shows this replication in action. A simple order-to-cash process in one company can not only be replicated in other companies in the value chain, but can also provide the clear interface points between companies: order placing and accepting, goods dispatch and receipt, and invoice to payment.

With a proper set of process protocols in place, noncore processes can be alternatively sourced, without the fear of interfacing problems or becoming locked into an unsatisfactory situation.

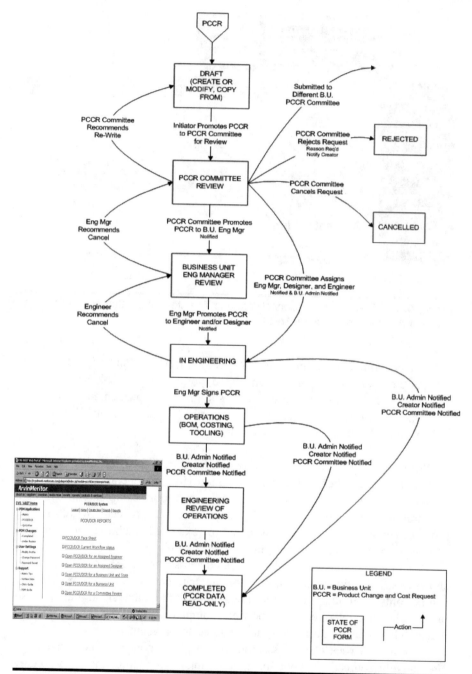

Figure 6.20. Steps 4 and 5: Consolidate and Transform the Enterprise (Source: ArvinMeritor™)

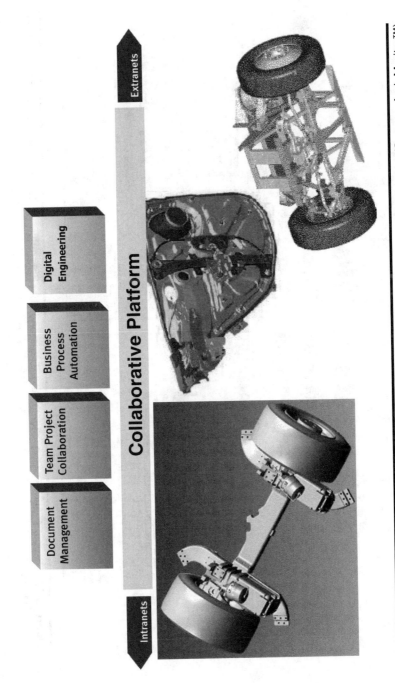

Figure 6.21. Steps 4 and 5: Consolidate and Transform the Enterprise — Digital Engineering (Source: ArvinMeritor™)

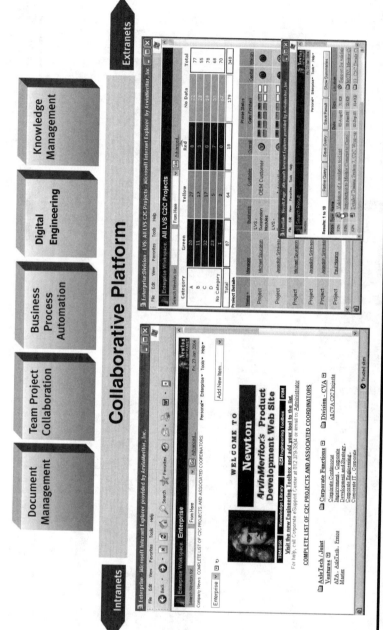

Figure 6.22. Steps 4 and 5: Consolidate and Transform the Enterprise — Knowledge Management (Source: ArvinMeritor™)

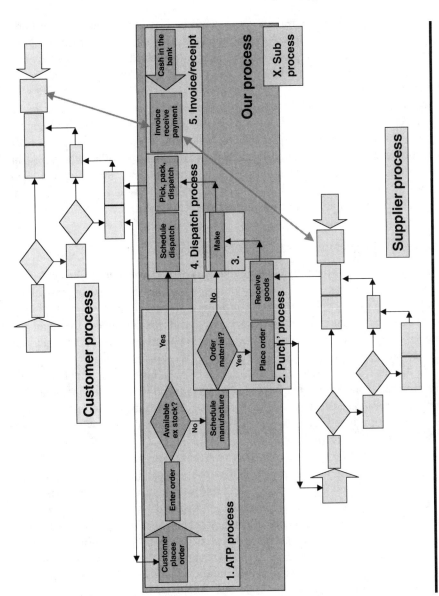

Figure 6.23. The Order-to-Cash Process

Organization and Location

Value chain organizations reflect cooperation and collaboration between partners. As we have seen, the collaborating organizations will put in place governance models that manage the shared affairs of the extended enterprise. The authors believe that these foundations will pave the way for a few value chains to achieve level 5.

Such governance models require the provision of performance data to be collected across the extended enterprise How else will they govern? Increasingly, this will become the province of automated processes, written under appropriate BPML protocols, which will automatically collect the relevant information from the various enterprise systems and present it in a collated form as and when needed. Again, the foundations of process management are seen to be an essential part of the inter-enterprise collaboration.

Data, Applications, and Technology

Increasing the level of supply chain maturity is driven by process. This is where the value is generated. This can only happen with the will of the organizations involved, and organization and location follow, to institutionalize and socialize the changes. The enablers that underpin this condition and make the management of complexity possible are data, applications, and technology, but they are enablers and increasingly will be invisible to the business users.

As the processes begin to flow across the boundaries, the extended chain will need real-time visibility of key partner data (such as stock dispositions and demand forecasts). The role of BPM and the architectures such as CSC e4^{SM} becomes paramount. Figure 6.24 shows the structure of CSC e4^{SM}. A number of key features make this ideal at level 4:

- All applications are plugged into the service layer with lightweight adaptors, making their data available to the rest of the enterprise.
- All the process rules are managed by the business process manager at the heart of the architecture, which sits between the applications and the user, who no longer needs to know with which applications she or he is working. Written in a process language that obeys the protocols discussed earlier, this process engine can now talk to other process engines in the value chain.
- Business users see the system through the coordination layer. They are presented by the business process manager with the information they need to execute their process, which then shares back the results with the appropriate applications. Of course, if the technology managers want

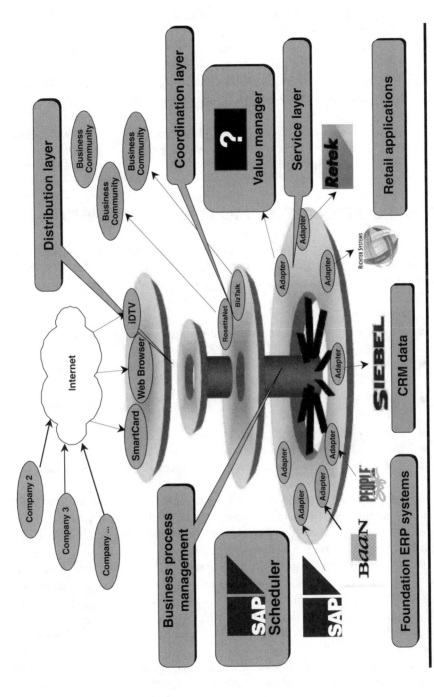

Figure 6.24. CSC e4ˢᴹ Architecture

to upgrade an application, they can now do this without the business users being aware. Similarly, if the business users want to change a process, as long as the data are available, they can do this without reference to the technology managers.

■ Finally, the extended enterprise talks to itself through the distribution layer. The use of browser-based interfaces means that all the different applications in the entire enterprise can share, under the control of the process engines. The underlying applications are transparent to the business users, who can concentrate on managing the whole value chain — the true glass pipeline!

TRANSITION FROM LEVEL 3 TO LEVEL 4: THE BENEFIT CASE

Because there are few true examples of the level 4 extended enterprise, value data on the benefits achieved are light. However, the results that have been achieved are significant and point the way for others.

The level 4 value chain is characterized by:

■ The whole chain is agile, lean, and resilient, with quality at the highest possible level.
■ Key processes flow up and down the chain between the enterprises, so that
■ The flow of key commodities is coordinated and, for those key commodities,
■ Costs and inventories have been minimized.
■ The value to all participants in the chain is increased and
■ The end customer is experiencing excellent service and satisfaction.

One of the best examples of a level 4 value chain is the one built by Dell. By implementing supply chain software from i2, Dell can now procure inventory from suppliers over the Web in real time and pull materials into the factories every two hours based on customer orders.

To meet the demands of these orders and to build the customized systems in a timely manner, Dell created efficiencies in its materials management system. With the implementation of i2's supply chain tools, Dell has enhanced its procurement processes so that almost 90% of the company's purchases from suppliers are on-line and there are only two hours of inventory on the factory floor.

A key component of Dell's supply chain management is having materials in close proximity to Dell factories; therefore, suppliers are required to have inventory hubs near the manufacturing plants. A huge benefit of this supply chain solution is communicating with these hubs in real time to deliver the required materials needed for every two hours of Dell customer orders.

Dell launched valuechain.dell.com, a secure extranet that acts as a portal for Dell suppliers to collaborate in managing the supply chain. This site offers suppliers a customized view of their materials at Dell, including reports on material quality, performance management scorecards, negotiated and forecasted cost reports, engineering change orders, supply/demand forecasting tools, and material demand.

"Our supply chain process was very well-executed, but we had an opportunity to increase our efficiency, make the process paperless and provide a single system of record that we and our suppliers could share," Eric Michlowitz, Dell's director of supply chain e-business solutions, is quoted as saying (source: http://www.dell.com/us/en/gen/casestudies/casestudy_dell_i2.htm).

CONCLUSIONS

The results of the authors' survey of over 40 case studies for the transition from level 3 to level 4 are summarized in Figure 6.25. Again, the overall percentage savings recorded were, in some cases, not as great as those for the earlier

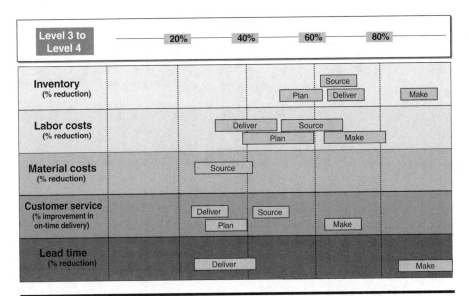

Figure 6.25. Moving Up the Supply Chain Maturity Model from Level 3 to Level 4

transitions, but, again, they build on those earlier gains, keeping level 4 companies and their extended value chain partners out there ahead of the competitive game.

But there is still a level to go — in the authors' view, still more gains to be made — by getting to level 5. Level 4 extended enterprises still suffer from the need to negotiate changes and improvements one by one. Often, there is a value chain lead company that "calls the shots," leaving doubt among the other participants that their value may not be increasing as fast as the leader's. The key question is whether the value of the whole chain increases when the leader benefits. We will find some of the answers in level 5.

7

ASPIRATIONS AND REALITIES OF LEVEL 5

This chapter explores the dreams and promises contained in making the highest progression in the maturity model — attaining a level 5 position, something very few firms have achieved. This final step is presently more theoretical and objective than actual, as few businesses comprehend the concepts and meaning behind full network connectivity — an environment in which business allies have linked themselves from end to end in a value chain so they can share vital knowledge in an instant and manage value in that chain as though they were a single enterprise. Most companies, moreover, are either studying the underlying ideas before making a test run or have found their organization unwilling to enter into the necessary electronic sharing of information and knowledge vital to attaining a real network environment. A few have the interest, but lack the technical capabilities necessary to link their organizations electronically to an extended enterprise.

For those businesses that do make the transition, a true competitive advantage is gained as the full network connectivity provides more firsthand, instant information on matters of importance, greater flexibility in creating and providing responses to demand, and the ability to use electronic communication to show visibility across the total network. Co-managed, intelligent value chains characterize level 5, with an overarching value proposition that makes sense for the value chain constituents and the targeted customers and end consumers. Indeed, there is evidence that those few organizations that have achieved such a level are able to take advantage of this intelligent value network, characterized by an ability to truly shift the emphasis to the top line and build new revenues in targeted areas better than competing networks. Investment decisions are shared as a matter of course in this type of environment. Processes flow seamlessly from end to end, and day-to-day planning and scheduling decisions take into account information from members across the extended enterprise.

At the heart of such an environment will be business *partnering* between very close and trusted business allies, focused on gaining mutual advantages in an industry or market. Under the right conditions, new and higher levels of performance can be achieved. As one senior executive explains, "Supply chain partnerships, properly managed, could add significant value. Business pressures today are so great — and are rising so rapidly — that even the most sophisticated businesses may be failing to realize the full potential of their current relationships and missing opportunities to build new ones" (Rudzki, 2004, pp. 44–45). How important is achievement of such a situation? And why should firms set aside their normal reluctance to share with external allies to reach this level? As Robert Rudzki explains, "Businesses can no longer afford not to partner where it makes sense, and they cannot afford to partner poorly" (Rudzki, 2004, p. 51).

THE MUTUAL SEARCH IS FOR HIGHER LEVEL IMPROVEMENTS

The basic question to be answered, as a firm is poised to move with the help of willing and trusted business allies into what can only be described as virgin territory, is: Why should we make the trip? For most firms, reaching level 4 and gaining the results we have cited is progress enough, and further ventures will seem impossible, too risky, or as involving too much alteration to the existing business culture to pursue. Since an extremely high level of trust is the necessary ante for participation, most firms will rule themselves out of the game. The idea that a major retailer, for example, would be willing to work closely with a few key suppliers to determine how they can find higher level savings, share the benefits, and work closely to draw more revenues to both parties is simply an alien concept in many business circles.

For those with the determination to reach this level, a simple illustration of the conditions that prevail is presented in Figure 7.1. Here we see that the various business applications reach the highest state in the maturity model. Design, development, and product/service introduction reach the point where the linked constituents in the value network come together and co-develop new introductions. Combining the expertise across several capable business allies, the focus is on joint design and development using a full business functional view, which provides any party that can contribute useful information to the effort a format from which to do so. Boeing typified this type of effort when

Progression / Business Application	Levels 1 & 2 — Internal Supply Chain Optimization (Supply Chain Optimization)	Level 3 — External Network Formation (Advanced Supply Chain Management)	Level 4 — Value Chain Constellation (e-Commerce)	Level 5 — Full Network Connectivity (e-Business)
Design, Development Product/Service Introduction	Internal Only	Selected External Assistance	Collaborative Design – Enterprise Integration and PIM-Linked CAD/CIM	Business Functional View – Joint Design and Development
Purchase, Procurement, Sourcing	Leverage Business Unit Volume	Leverage Full Network Through Aggregation	Key Supplier Assistance, Web-Based Sourcing	Network Sourcing Through Best Constituent
Marketing, Sales, Customer Service	Internally Developed Programs, Promotions	Customer-Focused, Data-Based Initiatives	Collaborative Development for Focused Consumer Base	Consumer Response System Across the Value Chain
Engineering, Planning, Scheduling, Manufacturing	MRP MRPII DRP	ERP – Internal Connectivity	Collaborative Network Planning – Best Asset Utilization	Full Network Business System Optimization Shared Processes and Systems
Logistics	Manufacturing Push – Inventory Intensive	Pull System Through Internal/External Providers	Best Constituent Provider	Total Network, Virtual Logistics Optimization
Customer Care	Customer Service Reaction	Focused Service – Call Centers	Segmented Response System, Customer Relationship Management	Matched Care – Customer Care Automation and Remediation
Human Resources	Regulatory Issues/Hiring, Recruiting, Training	New Work Models, Training	Inter-Enterprise Resource Utilization, Training	Full Network Alignment and Capability Provision
Information Technology	Point Solutions Internal Silos	Linked Intranets Corp Strategy/Architecture	Internet-Based Extranet Shared Capabilities	Full Network Comm. System Shared Architecture Planning

Figure 7.1. Five Levels Explain the Supply Chain Evolution (Copyright ©2000 E-Supply Chain: Using the Internet to Revolutionize Your Business)

that firm decided it could cut the normal production time on its new 7E7 aircraft from over a year to four months, by allowing key suppliers to take responsibility for major components of the plane. These suppliers not only provide key sub-assemblies, but were very active in the design of the aircraft and its development process.

Purchasing, procurement, and sourcing reach the point where network sourcing is done through the best constituent, which means that raw materials and services are secured from the best possible source (regardless of global location) through the member of the network most able to secure the best arrangement. Often, members use each other to manage specific categories of the buy, so the task is spread across more participants, and the best arrangements bubble to the top of the effort. In this advanced level, the normal emphasis on price negotiation is set aside so that the most strategic suppliers can contribute to the sourcing methodology in a way that benefits both buyer and seller.

As two writers describe the condition, "Supply chain managers need a sourcing methodology that will not only satisfy the near-term cost concerns but also position the organization for long-term operational success and profitability." Their idea is "to extend operational efficiency while not losing sight of the core competencies that sustain profitable growth." We contend that such a situation is achieved when collaboration between buyers and their best sources establishes a plan that has mutual advantage. Such an approach can be termed "predictive sourcing," under conditions where "the company's sourcing strategy is linked with its overall business strategy" (Bernhard and Vittori, 2004, p. 58). We would add that the linkage should be across network strategies, requiring the network participants to share information on the best sourcing strategy, resulting in a joint business plan with industry-best objectives as the end result.

Kyocera Wireless, the Japanese manufacturer of telecommunications products, offers a good example at its San Diego operations facility. This firm decided to source manufacturing in the United States based on an assessment "of where the company adds value for its customers." For Kyocera, predictive sourcing meant "looking at manufacturing, logistics, and demand planning needs collectively — instead of basing sourcing decisions only on manufacturing costs" (Bernhard and Vittori, 2004, p. 61).

Marketing, sales, and customer service also reach a higher plateau, where an electronic consumer response system is in effect across the full value chain. Now the targeted best consumer groups are contacted directly and provided customized services through a variety of channels: store, catalogue, telephone, Internet, or personalized delivery. The movement results in what is being called the "demand-driven supply network" (DDSN). According to AMR Research, such a condition results in a "system of coordinated technologies and processes that senses and reacts to real-time demand signals across a network of customers, suppliers, and employees. It enables organizations to improve operational efficiency, streamline new product development and launch, and maximize margin" (Asgekar, 2004, p. 15). We cite DDSN as part of our roadmap to the highest level supply chain processing, where the demand from the most important customers is linked to the capability to supply so that optimum conditions can be achieved across the full network.

Engineering, planning, scheduling, and manufacturing move to a very special area, beyond Six Sigma quality, beyond enterprise resource planning sharing, further than cost containment and manufacturing excellence, and into the realm of total enterprise optimization. Balancing supply and demand will be facilitated through the sharing of current consumption data and the capability of the network to respond and deliver, essentially harmonizing sell-side and buy-side activities into a composite system of response. Through the sharing of optimum processes and enabling systems, the value network is hard at work turning out the best products at absolute lowest feasible costs in a manner that reaches highest possible productivity efficiency at each manufacturing site. Asset utilization moves to unprecedented levels, as machines and equipment are only used if they perform at industry best standards. In an allied vein, logistics becomes a total network effort, and a state of virtual logistics optimization is achieved, in which the participants are attaining maximum storage and transportation efficiency. The key is the electronic visibility and use of virtual systems that allow any constituent to find the best possible warehouse and delivery mechanism for the products provided and the destination involved.

Customer care becomes a special area, in which the services provided to the customers and consumers are matched with actual needs. The process is automated for those who prefer little human interaction and like to find electronic solutions to their needs. Human contact of the highest caliber is provided at the other end of the spectrum, for those who want the same person handling their issues in a professional manner. In between, service and remediation are provided, better than by any competing network, because of the access the participants have to on-line, accurate knowledge pertinent to the issues being discussed.

Human resources now has a full network alignment, which means the focus is on helping the members find the best skills and individual talents, through a sharing of such knowledge, and the actual sharing of resources between the companies involved.

Information technology rounds out the capability matrix, as full network connectivity is provided for all the key players. Business intelligence will be provided to all parties with a need to extract data from collective databases. Retailers, for example, will be enabled to sort through their enormous number of transactions to find the trends that need most attention and where they should go for help from their manufacturers and suppliers. They will share this knowledge upstream in their supply chain, as trusted allies further analyze the data and add valuable insights while responding in industry-best cycle times. Through this type of electronic window, any important constituent will be able to immediately link into the processing and view what is transpiring. A shoe manufacturer, for example, will be able to search its network, from the beginning materials (leather, synthetics, and dyes) to the intermediate manufacturing steps, through all delivery options and in-transit movement, to the final retail stores and consumption by a particular consumer. Returns will be included, as well as all repacking and reshipments.

In the middle of this advanced information technology environment will be the integration of information systems, business processing, and applications. Through the sharing of ideas on the right systems architecture and business process management systems, this state of network connectivity will allow the inclusion of other firms of any size for any particular product. The parties will be able to go into the collective databases and remove vital knowledge from components that offer quick access to information that would take competing systems days to accomplish.

CO-MANAGING VALUE: THE ESSENTIAL LEVEL 5 DIFFERENTIATOR

Moving from level 4 to level 5 in the extended enterprise is analogous to moving from level 1 to level 2 in the stand-alone company. Figure 7.2 summarizes the differences between levels 1 and 4. At level 4, while the level of collaboration is high, any substantive changes needed to improve performance still have to be negotiated on an item-by-item basis. They are always weighed against the impact on local shareholder value, often with little regard for the overall impact on value in the extended enterprise.

The authors recall a conversation with the logistics manager of one of the U.K.'s major retailers. The retailer had just announced that its distribution centers were about to go stockless. When questioned about how this had been achieved, the manager said that suppliers would deliver more frequently, in smaller lots, closely matched to the most recent store demand data. Later in the conversation, the question of value arose. Had the value in the chain overall gone up or down? Not only did the manager not know, but the question had not occurred to him. If the value had been reduced, at the suppliers' expense of the extra less efficient deliveries, then this would eventually find its way back into the value chain's economics. The suppliers would gang up on the retailer to force more profitable dealings, they would fail as their profits were reduced,

Level 1 is characterized by:	Level 4 is characterized by:
• Local, departmental performance measures	• Company performance measures, driven by local shareholder value
• Processes with many handoffs between departments	• Processes with many handoffs between companies
• Multiple views of the same demand data	• Multiple views of the same demand data
• Buffering with inventory or capacity to compensate	• Buffering with inventory or capacity to compensate
• Investment driven by local, departmental goals	• Investment driven by local, company shareholder goals

Figure 7.2.   The Analogy Between Level 1 and Level 4 Transitions

or at best, they would compensate for reduced profit by cutting back on product development or other areas of customer service.

So how should a level 5 extended enterprise manage value? We suggest one approach, based on an example of the value chain to deliver a bottle of carbonated soft drink to the end customer. The supply chain for this simple product is surprisingly complex, from plastic at the upstream end through the bottling and packaging processes to the supermarket shelf. Even two or three years ago, we would not have contemplated treating this as a single entity. For one thing, we did not have the necessary tools. Now, the Internet and effective process management mean we can move information to where we need it, and fast schedulers give us the power to synchronize the flow of materials through the whole network.

The challenge we face is bringing it to life, winning the extra value and sharing it equitably between the partners. And that requires doing no more than we have done many times before at level 1 — except that now the "functions" are the businesses in the extended supply chain. Mapping the chain is no different than mapping a level 1 supply chain, except that now several companies are involved. The map is shown as Figure 7.3.

Given that we can map the extended enterprise, we can also create an economic model for it, a pseudo-profit-and-loss statement and balance sheet. By combining these, we can build an economic value added (EVA) model, similar to and with the same role as the level 1 return on net assets model, with the exception that, for the extended enterprise, we must factor in the cost of capital for each of the partners. An example is show as Figure 7.4.

For the extended chain, we must also build EVA models at two levels:

- For the chain as a whole
- For each key player

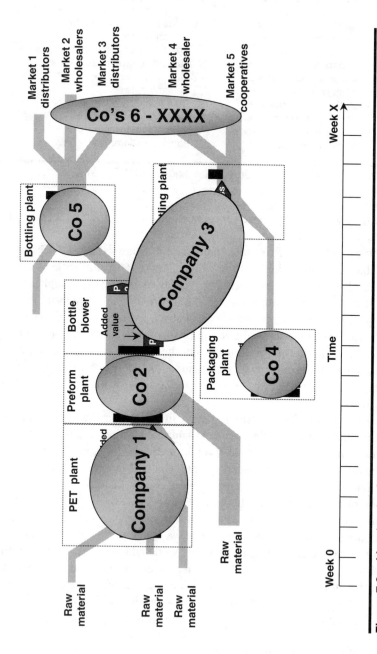

Figure 7.3. Mapping the Chain: A Schematic Supply Chain Map for a Plastic Drink Bottle May Look Something Like This

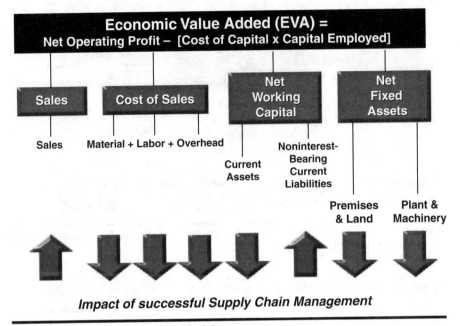

Impact of successful Supply Chain Management

Figure 7.4. Creating the EVA Models

We do this so we can see where value is gained and lost across the chain, so as to manage it fairly. This is analogous to the investment in the active ingredient plant in the Zeneca example in Chapter 3, where the value was "gained" elsewhere in the chain.

Managing value at level 5 relies on the power of the maps and EVA models. First, transactions in the chain can work much more efficiently. There is no more "buying and selling" — the extended enterprise analogue of transfer pricing at level 1! The flow of information and cash is shown in Figure 7.5.

The base case, to start the enterprise trading in this new way, would replicate the "as-is" condition for price and margin. Extra value would then be divided up, according to an agreed formula, to reward improvements across the whole chain. Of course, governance in the extended enterprise will become critical to success, much as it is with a level 1 board of directors. In particular, there will have to be strict rules for managing changes in the enterprise (e.g., deciding how partners get voted off and on).

More interestingly, with a value management mechanism covering the whole extended enterprise, managing the investment opportunity in an ongoing way becomes a real possibility. We are no longer locked into looking at each investment case for its impact on individual companies in the chain to ensure their local shareholders are protected.

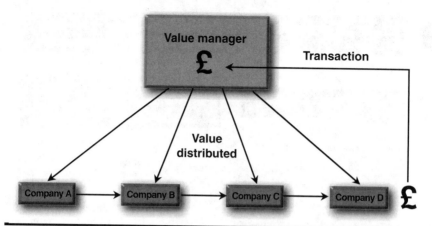

Figure 7.5.   Value Management in the Level 5 Enterprise: The Transaction Process

By understanding where investments are made and the resulting value created, as illustrated in Figure 7.6, we can create an environment where companies
are not penalized for investing for the good of the whole chain. Instead, an
investment is judged on the basis of the value it will create in the whole
network, and the risk of the investment is shared by the value chain partners.
When the investment creates the value projected, the investing company is
"reimbursed" and the extra value shared.

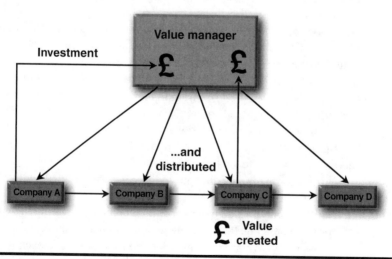

Figure 7.6.   Value Management in the Level 5 Enterprise: The Investment Process

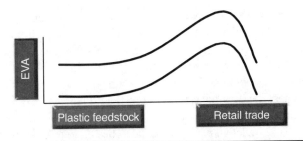

Figure 7.7. Unlocking the Opportunities (Copyright ©2001 CSC. All rights reserved.)

Of course, the trailblazers will need to solve many practical issues. For example, we may need two categories of investment: one for the good of the whole chain, with benefits to be shared, and one for an individual partner investing for local benefit, with that benefit to be retained. Governance must watch for market shifts; in our example, perhaps that would be a major industry-wide change in bottling technology, resulting in major cost reductions. The partner in that part of the chain would have to share the benefits of local investment or the chain as a whole would become uncompetitive. If the partner is not willing to share, the governance body has the ultimate sanction of replacing the partner!

A little research into most value chains will uncover many value-creating opportunities. The interesting thing is that an understanding of the value chain and its value models will show that EVA is not equally spread. In the case of our soft drink bottle, it is heavily skewed to the downstream, as shown in Figure 7.7. No proposition will succeed if it relies on persuading one player to give up part of its value. But it just might succeed if, by collaborating, we raise the whole EVA curve and generate more value for everyone!

Ultimately, the level 5 extended enterprise will outstrip its lower level competitors in cost, quality, and time to market. The entry ticket is a willingness to engage, and the prize is a share in the increased value in the entire extended enterprise.

THE LEVEL 5 EXTENDED ENTERPRISE: A BLUEPRINT FOR BUILDING CUSTOMER VALUES

As supply chain efforts mature, and a few companies do move to the most advanced levels of performance, one distinction becomes apparent. Firms that embrace the inherent concepts, as part of a total extended enterprise optimization effort, have gained the high ground. The leaders have used advanced techniques to focus first on operational excellence and then on customer satisfaction, to open a serious gap between less able competitors, with several companies beginning to dominate their industries. Positions achieved by such leaders as Wal-Mart, Procter & Gamble, Toyota, Intel, Nike, and Dell bear witness to the values being added through a concerted enterprise-wide improvement effort. These leaders have discovered the advantages offered by moving their supply chains into a position of having superior capabilities, gained through greater access to knowledge across what becomes an enterprise-wide intelligent value chain network.

When that knowledge is combined with an effort to develop greater customer intimacy and satisfaction, especially with the most important customers, the advantage becomes an ultimate distinction in most industries and markets. To complete this chapter, we want to explore the advantages to be gained by applying all of the premises presented earlier in this text to a business system that operates in an intelligent network environment — a true level 5 operation. To establish the goal of such a network, we would cite a statement made by Debra Hofman of AMR Research: "Companies need to optimize the demand and supply side of the network so there is a transparent flow of information that brings the components together into a finely tuned network that is synchronized, high performing, agile and responsive to changes in the signal" (Hofman, July 2003). We could not agree more, and when these conditions are attained, a market advantage is the reward.

HIGHER GROUND CAN BE GAINED

A recent survey by CSC, in conjunction with *Supply Chain Management Review* magazine (Quinn, 2004), clearly documented that savings and improvements are real for serious supply chain efforts, often reaching three to eight points of new profits. This study, as well as ones conducted by AMR Research and other major consultancies, also shows that the important savings (particularly those related to revenue increase) are eluding most firms, which are still bogged down in the early levels of the maturity and SCOR® models and not inclined to work with external business partners.

We see an enormous possibility in such a context. The opportunity to use supply chain as a driving force behind further performance enhancement, and to move a firm into a position where the distinguishing feature is being solidly linked in an intelligent value network, has become the means to reap the greatest return from an end-to-end supply chain improvement effort. Internal obstacles and cultural conflicts tend to be the greatest inhibitors to achieving such a position, the most important of which are good data management and overcoming process difficulties.

Distancing an individual business from its competitors in areas of importance in a market has long been the goal of most enterprises. The chance to extend market leadership, however, and to gain a dominant position through the application of collaboration and technology focused on customer satisfaction, the key ingredients of the intelligent value network, is not so well known and has never been greater — for those businesses willing to overcome normal cultural barriers and the traditional unwillingness to work cooperatively with external resources to cope with process problems.

This opportunity is achieved by linking together four topics of importance to today's businesses: supply chain management, customer relationship management, technology application, and customer intelligence. The last topic is our terminology for the acquisition, management, and integration of customer knowledge in order to create a differentiating customer value proposition for the whole extended enterprise. By looking holistically at these usually disparate topics, companies can develop integrated strategies and solutions for delivering products and services to key customers better than any competitors. When the effort is extended through business process management techniques to include willing and trusted business allies working across an extended enterprise for the same purposes, the advantages are unmatched.

IMPROVEMENT STARTS IN A COMPLICATED ENVIRONMENT

Any discussion on the possibilities of achieving total enterprise optimization begins with an understanding of just how complex an extended enterprise supply chain has become. While the original supply chain efforts were directed toward achieving optimum operating conditions across a linear set of tightly linked, internal process steps, from beginning raw materials to final delivery of products and services, Figure 7.8 reminds us that most supply chains are becoming complex business systems. Any analysis that is limited to internal processing is doomed to operate with suboptimized conditions. There are simply too many players in a typical business network, most of which are global in extent. The end-to-end processing that has come under scrutiny for improvement now includes a multitude of business partners. Concurrently, the necessary flow of information and knowledge within a business network has become as important as the physical flow of goods and the transfer of money across what is clearly an extended enterprise. Supply chain optimization now requires the collaboration of a host of business partners working in concert for the same end results.

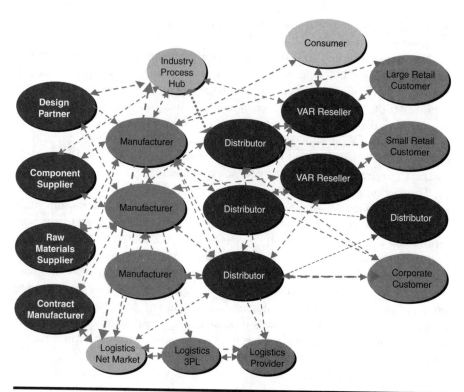

Figure 7.8.   Supply Chains Are Becoming Collaborative Networks

It becomes imperative in such an environment that the firm seeking optimized conditions make a passage from an internal-only perspective, in terms of generating process improvement, to one where willing and trusted business allies are made a part of the process improvement effort, with the end result focused on customer satisfaction. To accomplish this objective, the leading firms are merging their advanced supply chain management concepts with their customer relationship management efforts, yielding a framework and roadmap for progressing through a series of levels until the highest possible return on the effort, in terms of value for the customer and benefits for the providing firm and its allies, is attained. Along the way, a concurrent effort must be made to balance supply chain progress with the firm's customer relationship management capabilities and synchronize the results of the two efforts. In the next chapter, we will delve into these necessities, as we explain how firms partnering in level 5 can take advantage of their intelligence sharing and outdistance competitors in the creation of new revenues.

TRANSITION FROM LEVEL 4 TO LEVEL 5: THE BENEFIT CASE

Level 5 becomes "the final frontier" — unlocking the ultimate supply chain benefits. With diligence and true partnering efforts, the level 5 extended enterprise will have the following characteristics:

- The whole supply chain is agile, lean, and resilient, now operating end-to-end in the same way as a level 2 business operates internally.
- Costs are minimized.
- Inventory is minimized.
- Value to all participants in the chain is maximized.
- The end customer is experiencing excellent service.

YANTRA STUDY POINTS TO THE POTENTIAL IMPROVEMENTS

We now return to the Yantra/CellStar case study presented earlier, to emphasize how these firms found real improvement through advanced supply chain management network-type efforts. One of the most important factors in the effort to improve the delivery system for the intended wireless customers was the ability to handle reverse logistics. CellStar had determined that this factor was the Achilles' heel in the wireless industry. Traditional reverse solutions, for example, could not address the need to:

- Maximize value of the returned handsets
- Track handsets across the enterprise, including the repair process
- Evaluate individual handset attributes (software, warranty, etc.)
- Reconcile issues with multiple vendors and distributors
- Manage multiple repair vendors
- Assess product disposition at the unit level

Before developing its Omnigistics solution with Yantra, CellStar had little chance of providing answers to these problems. Its returns management system had no process-wide status. Reporting lacked a common format and was not capable of handling details at a unit level. There was no ability to track unit status (disposition decisions were made at the pallet level) and no mechanism to track or manage warranty claims. In short, it was operating with a slow, costly, and inflexible return-handling system.

With the help of its business ally, CellStar introduced Omnigistics and brought a new and better dimension to reverse logistics services. In addition to the software development cost, a multimillion-dollar investment was made in business process reengineering the existing processes. With the help of Yantra as the provider of the new and complex supply chain management system and CSC as the systems integrator, CellStar also established a 230,000-square-feet dedicated facility with specially trained forward and reverse logistics teams. This facility became a certified repair center, with a dedicated account management team.

Facing the issue of reverse logistics, the team adopted level 5 principles and developed a new strategy and solution based on:

- **Collaboration** — To integrate the customer's business system, key business partners, and CellStar's logistics operations in order to provide help from a seamless entity
- **Visibility** — To provide electronic access into the processing at both the enterprise and unit level
- **Speed** — To compress event cycles

The solution delivered a single, seamless management system that streamlined operations across the end-to-end reverse logistics process. Now the reverse management system was centralized and could process at the unit level. Reporting was on demand at the enterprise and unit level. Inventory management decreased the amount of nonearning assets and maximized the high turnover stock, while reducing or eliminating the slowest moving items. Asset reclamation options were included in that part of the system. Multiple repair centers were linked so they could monitor status at the unit level. Finally, multiple warranty programs were in place to manage and track claims.

To illustrate the advantages, a major national retailer became an early customer of the enhanced system. This retailer's handsets were returned at the store level and replaced with new units from store inventory. The returned handsets were sent to the retailer's centralized return center and then to a carrier for disposition. The issues were symptomatic of the industry at the time. The process was marked by escalating costs and loss of profitability. The value of the returned handsets to the carrier was not known, and returns were difficult and costly to handle at the store level.

With the new solution, all returned handsets were forwarded to CellStar for disposition. The carrier handled the process claim, while a specially selected broker sold the assets at the most efficient rate. Disposal was based on industry and environmental standards. The time to reconcile warranty credit was reduced from over 30 days to less than 8 days. The average cost recovered through brokerage of returned units was $90. The total time to process the returns went from over 45 days to less than 10 days. Additionally, the returns disposition analysis identified 6,000 out-of-warranty handsets that qualified for manufacturer's product recall, yielding a $600,000 credit to the retailer.

CONCLUSIONS

As mentioned, level 5 is rarified space. The notion of full network connectivity is still too new and too much in opposition to the normal business approach to sharing of data and knowledge. When a dedicated set of trusting business partners determined to link a portion of their information system, and use business process management to transfer knowledge that helps each other, a discovery is made. It is possible for firms of any size and character to participate in the creation of intelligence that eludes most businesses and establish a new route to profitable revenues. The remaining ingredient in that collaboration is bringing a focus from the analysis and data sharing to the customers and consumers of choice, a subject to be pursued as we study how the intelligent value network proceeds to use the advantages of business partnering to build the top line of the profit and loss statement.

8

CREATING THE INTELLIGENT VALUE NETWORK: A BLUEPRINT FOR LEVEL 5

To reap the most benefit from level 5 efforts, the linked businesses should apply their efforts toward specific customers and consumer groups, such that the perception these groups have of the network is one of superior capabilities and the one that renders the greatest satisfaction — or, more importantly, the greatest overall value. Attaining such a condition requires the features of supply chain maturity to be matched with a "customer intelligence progression." That is, the network allies will be using the knowledge being shared, as well as the process improvements, to distinguish the final results in the eyes of the most important buyers.

Using the maturity model, which is repeated as Figure 8.1, to describe the progression of supply chain efforts, we are reminded that the first two levels are internal only, where focus is brought to functional improvement and operational excellence to internal operations. The cultural wall standing between levels 2 and 3 represents all of the collective inhibitions and obstacles to accepting an external view of the processing and working collaboratively with willing business allies to build network improvements, which distinguish the value chain in the eyes of the most important customers. Levels 3 and 4 rep-

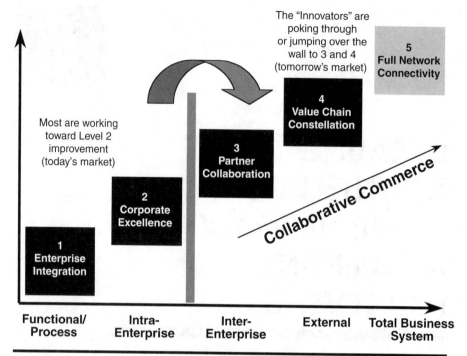

Figure 8.1. Supply Chain Maturity Model: Evolution to Industry Networks and Collaborative Commerce Is Inhibited by a Lack of Customer-Focused Processes and Organizational Alignment

resent the positions achieved by market leaders, while level 5 is intended to indicate the presence of total network connectivity with the highest processing capabilities.

As Figure 8.2 illustrates, customer relationship management (CRM) progress can be matched with the normal supply chain evolutionary levels, as a business moves from the early points of enterprise integration (where the firm gets its house in order), to a position of analytical CRM (where the firm becomes a viable part of a superior value chain constellation), and on to development of the intelligent CRM position (where the firm and its allies use superior knowledge and shared insights to dominate an industry). Alex Black, partner at CSC, was extremely helpful in the development of this analysis, and we express our thanks to him.

Beginning in the mid-1990s, most firms progressed through the first levels of the supply chain evolution, moving from enterprise integration, where early savings were made through concentrated sourcing and logistics efforts, to corporate excellence, where internal obstacles were conquered and planning, order

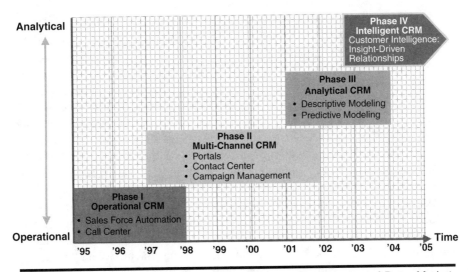

Figure 8.2. Customer Intelligence Is Driven by the Convergence of Data, Marketing, and CRM Applications Capabilities

management, manufacturing skills, and inventory management became serious parts of the effort. During this time, many companies also progressed into a form of operational CRM. Sales force automation became a factor, as companies learned they could use customer data to enhance the ability of sales representatives to help customers find extra values and build more revenues. Call centers came into vogue as contact centers were established to match the services needed with what would truly help the key customers and to guide responses to customer needs through multichannel customer service hubs. Toward the end of that period, while in the second phase of the effort, campaign management became a factor, as firms learned they could ally themselves with key suppliers and customers to improve the results of special sales efforts.

At the beginning of the new century, those firms that maintained a dedication to the supply chain effort moved into level 3, and began collaborating in earnest with their key business partners, to find the hidden values in the supply chain linkage that eluded those firms bogged down in an internal-only focus. During this period, these firms typically advanced to a form of collaborative CRM, applying technology to increase the knowledge available to business allies having the same purposes. Using the Internet as the major tool of communication, these companies began to share valuable customer and consumer information with selected and trusted business allies, so they could further improve their abilities to create and sustain new revenues. Partner relationship management became the tool of choice, as these allies learned they could share previously sacrosanct and private information to analyze trends and build revenues together, without risking the future of their organizations. Customer data integration became a vital technique to assemble and use important knowledge on customers, consumers, and markets to introduce customized solutions and offerings that were clearly better than any competing business network.

As a few businesses managed to progress into level 4 and became part of a value chain constellation, the more advanced firms moved further with analytical CRM and began to reap the benefits of a true customer intelligence environment. Here the nucleus firm in the center of the supply chain network would join forces with key supply chain partners and drive these network allies to analyze customer knowledge together. Using business process management (BPM) as the linking tool and BPM system as the linking system, to transfer the important knowledge between disparate communication systems, these firms found the means to quickly transfer valuable data among supply chain partners. One important output became demand chain management, where the actual needs of the end consumers and customers were matched with the capability to meet those needs. Essentially, demand chain and supply chain converged, and the intelligent value networks which emerged from this level of progress were best able to respond to what the market truly wanted in the most effective manner.

The requirements supporting this evolution are not exactly novel. Improving profitable revenues with targeted customers and retaining their loyalty have been central tenets of business strategy for a long time. With access to helpful knowledge buried in the burgeoning databases most businesses are building, it becomes a modern art, enhanced through technology applications. When the effort is extended to integrating CRM systems with advanced supply chain management (ASCM) efforts and access to customer intelligence, a chance to differentiate a firm and its closest business allies arises. Unfortunately, after almost two decades of trying, the concept of applying CRM in an enhanced supply chain to create new business is well understood, but the practical application appears to be very limited. Most firms are simply too afraid to share customer data with external sources, even when the objective is mutual benefit through increased revenues. The opportunity to make greater use of this capability looms as one of the most important challenges facing business today.

CUSTOMER RELATIONSHIP MANAGEMENT: A CONTEMPORARY VIEW

An analysis of the current state of CRM reveals that most markets are under serious scrutiny to show actual value for the necessary investments — in time, resources, and capital. Because of the many stories related to inadequate returns for the investment, CRM suffers from a poor reputation, in spite of the many successes that have been recorded. There is a high degree of complexity associated with these efforts and a naturally high cost of integration across an organization and its end-to-end network. As a result, current views of the potential values are tempered by a need to bring focus to immediate process improvement and bottom line returns. Nevertheless, when executed as part of a deployment of strategies, with enhanced processes and enabling technology applications that are used to acquire, develop, and retain an organization's best customers, CRM becomes a powerful tool for increasing revenue and profit.

In essence, a contemporary CRM operating model will serve to improve the characteristics and performance of a customer-intimate organization. The inherent characteristics for customer-intimate organizations will include:

- Creation of the best business solutions for the key customers
- Introduction of customized products and services to meet these customers' unique needs
- Presentation of a unique range of superior services, so customers can get the most value from the products delivered
- Establishment of the most flexible and responsive system of supply and delivery possible with current technology

The operating model benchmarks will include:

- Management systems geared toward creating superior results for carefully selected strategic customers
- A culture that embraces specific rather than general customer solutions and thrives on deep and lasting relationships
- Deep customer knowledge and breakthrough insights about the customer's underlying processes
- Decision making delegated to employees close to the customer (Treacy and Wiersema, 1995)

Reaching these conditions requires a lot of concerted effort and nurturing a cultural imperative that is often hard for firms accustomed to working within an internal-only focus. CRM has its roots in the idea that as a firm's supply chain moves toward maturity, it becomes more effective at both internal and external processing; that is, it improves its ability to process within its four walls, and then extends its learning, with the help of useful business allies, to constructing a network of delivery that has superior features from the viewpoint of the most important customers and consumers. Such an accomplishment means that CRM progresses from making the most of the best marketing and sales practices necessary to achieve parity or better against competing firms to the point where it is engaged in advanced techniques of value to the customer that are not found in any competing group. Since most of these techniques will require enabling technology and the sharing of vital knowledge with key external resources, Figure 8.3 depicts some of the features required in an intelligent value chain, which becomes the end product of a successful ASCM/CRM technology-enabled effort.

Within the intelligent value chain, business allies are working together from a right-to-left perspective. They begin with what it takes to have a competitively advantaged value network in the eyes of the most important end customers or consumers, and then they work backwards toward what the upstream side of the value chain should be doing across the enterprise processes to achieve the desired superior conditions. Together, the linked parties are working to find the best solutions and practices for all of the key process steps. Beginning with improved forecasting and moving through the necessary linked processes, the network partners apply their best resources to find greater results with product development and introduction, the ultimate distribution efficiency, the best methods for product replenishment, jointly developed marketing strategies, and the best possible order fulfillment system. Along the way, they work collaboratively to find the best enterprise processes and become extremely effective at any point of handoff between supply chain constituents. In short, they are working in concert to develop business in a manner that greatly satisfies the key customers and enhances profitability for all of the contributing allies.

Two requirements must be met as this intelligent value chain is constructed and nurtured. First, each participant or major constituent of what becomes the network of delivery must have attained a high level of capability in the supply chain maturity model (level 3 or beyond), an important element of which will be the ability to use BPM and its enabling business language, BPML, to enter and access parts of disparate databases so valuable knowledge can be extracted without compromising the security of the various systems. Second, the enabling technology applications must be selected collaboratively and be functioning successfully across the end-to-end network processing. That means the collaborating business allies are working in concert, with each making valuable contributions toward finding the enhanced state in which ASCM and CRM converge to create the desired differentiation in the eyes of the most coveted customers. They are doing this with the help of enabling BPM technology and superior systems across the end-to-end processing linking them into an intelligent value network.

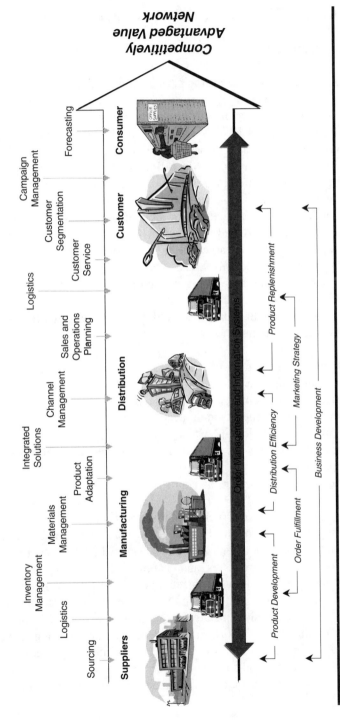

Figure 8.3.   The Intelligent Value Network: Building Network Business

THE VALUE OF CUSTOMER INTELLIGENCE

There is an important purpose behind the effort to establish greater customer intelligence. Bringing together a single view of the customer with high-value analytics can serve to optimize customer interactions, reduce operational costs, and enhance revenue-generating opportunities. To begin, most organizations have multiple records and accounting for the same customer, with no consistent information transfer across business units within the same organization. This condition leads to the absence of a single view of the customer and leads to inconsistent customer experiences. Much time and effort are wasted collating reports and gathering information, rather than focusing valuable resources on analyzing high-value information and knowledge. Much of the marketing effort, which is intended to build a demand, is focused on mass-market techniques, rather than the preferred targeted segments that offer the most lucrative returns on the effort. The inability to target the right customer at the right time, with no predictive modeling capabilities, exacerbates the problem and leads to ex-pending corporate energies on low- versus high-level customers and a total lack of optimized service levels.

Solutions to these complications can add dramatically to the firm's perfor-mance, including such features as:

- Data management personnel savings
- Faster call handling of inbound inquiries
- Prospect and customer solicitation savings
- Reduction in returned communications
- Improved data quality in critical operational systems
- Improved targeting for cross-sell, up-sell, retention, and acquisition campaigns
- Lower customer attrition or churn rates

More importantly, attaining such conditions puts the internal house in order and brings the firm to the point of being able to approach customer intelligence in a more contemporary manner. By today's standards, CRM has become the deployment of strategies, processes, and enabling technologies that are used to acquire, develop, and retain an organization's best customers. It includes un-derstanding customer needs, the relative importance of each customer segment, and the best, most economical means to meet those needs. Within an environ-ment focused on this view of CRM, strategy, processes, organization, and culture begin to revolve around a central focus dedicated to satisfying customers in the most appropriate manner and sustaining those with most strategic value indefi-nitely. Businesses adopting such an environment recognize that performing the end-to-end process steps in the most effective manner becomes the hallmark of network distinction, but it cannot be achieved without the knowledge necessary to optimize the important process steps.

Process orientation has never had more meaning in this environment. Organizations that remain internally fragmented and operate in a stovepipe manner will never achieve the advantages cited. They will be doomed to local optimizations within some business units and be prevented from achieving network process and systems optimization. Such systems as enterprise resource planning, CRM, and collaborative planning, forecasting, and replenishment simply will never be achieved in an optimal manner due to the process inefficiencies that will occur. Process design and enablement with new technologies and methodologies and tools are what will provide the greatest opportunity to increase corporate performance in the modern era. The drivers behind this return to a process focus, moreover, will be an enhanced customer-controlled environment, where customer satisfaction is the real end objective, with use of the Internet to create and control the sharing of valuable knowledge.

When ASCM and CRM converge in this advanced level of the evolution, some important characteristics will be apparent:

- Demand management and forecasting will be at improved levels, with actual need matched with capability to supply.
- Sales and operations planning will move to advanced planning and scheduling, where key suppliers and customers participate in diagnostics and planning sessions to bring a reality to the planning and supply processing.
- Inventory management will be a network effort, in which the linked allies work to deliver the right goods to the point of need in the right quantities at the right time.
- Visibility into the end-to-end processing will be on-line, real time, allowing the constituents to view what is taking place, track important events, and adapt the supply chain to ever-changing market conditions faster and more accurately than the competition.
- Event management will be at the highest possible level of effectiveness, as the reactions to any planned sales effort will be instantly relayed back to the important upstream partners, so they can react appropriately to actual event conditions and results.
- Investment in the extended enterprise will be driven by the good of the whole network, not just individual partners' local shareholder needs.

In short, the voice of the customer will be driving the supply/value chain, based on the segmentation that has determined the level of response necessary to satisfy the customers being served. This condition requires the firm to move

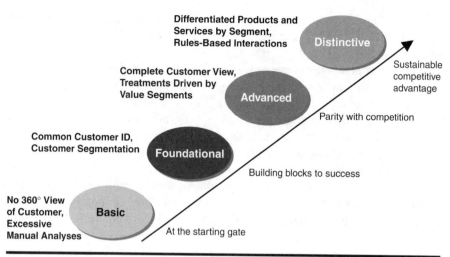

Figure 8.4. Customer Intelligence Maturity Model

through the four levels of the intelligent customer maturity model, as illustrated in Figure 8.4.

Beginning in the first level, or the starting gate, the basic requirements must be met. The firm adopts a 360-degree or total view of the customer and deals with the fact that most processing involves excessive manual analysis and handling of the data. Use of the intranet, or internal communication system, has to be improved so there are no cultural inhibitors to building the most accurate, accessible knowledge on the most strategic customers.

In the second foundational level, where the firm begins to erect the building blocks to success, a common customer identification system is installed and customer segmentation is used to separate the customer list by strategic value to the firm and profit to the firm. The matrix presented in Figure 8.5 has been used as a guide for such segmentation. It progresses from the low-value, low-profit, or "usual suspects" to the high-value strategic customers that are "to die for." Within each block of the grid are comments intended to help the selection process.

Two types of analytics are then used to identify and target the highest potential customers. A profiling tool is used to determine who and where the best customers are and what they really need. In this "descriptive modeling" area, focus is brought to such elements as lifetime value, demographics, behavioral trends, and decile analysis. A targeting tool is used to determine how the firm identifies the right offer to the right customer at the right time. In this "predictive modeling" area, focus goes to propensity to churn, chances for cross-selling and up-selling, and the propensity to buy.

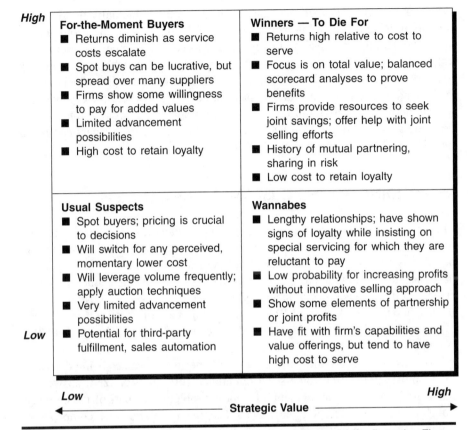

For-the-Moment Buyers ■ Returns diminish as service costs escalate ■ Spot buys can be lucrative, but spread over many suppliers ■ Firms show some willingness to pay for added values ■ Limited advancement possibilities ■ High cost to retain loyalty	**Winners — To Die For** ■ Returns high relative to cost to serve ■ Focus is on total value; balanced scorecard analyses to prove benefits ■ Firms provide resources to seek joint savings; offer help with joint selling efforts ■ History of mutual partnering, sharing in risk ■ Low cost to retain loyalty
Usual Suspects ■ Spot buyers; pricing is crucial to decisions ■ Will switch for any perceived, momentary lower cost ■ Will leverage volume frequently; apply auction techniques ■ Very limited advancement possibilities ■ Potential for third-party fulfillment, sales automation	**Wannabes** ■ Lengthy relationships; have shown signs of loyalty while insisting on special servicing for which they are reluctant to pay ■ Low probability for increasing profits without innovative selling approach ■ Show some elements of partnership or joint profits ■ Have fit with firm's capabilities and value offerings, but tend to have high cost to serve

High (top left), *Low* (bottom left)

Low ← **Strategic Value** → *High*

Figure 8.5. Customer Segmentation by Strategic Value and Profit to the Firm

Returning to the maturity model framework (Figure 8.4), as a firm moves into the "advanced" level 3 of the progression, it begins to build a market advantage, through ASCM/CRM. Now the organization is working selectively with business allies and develops a complete customer view, with treatments and services matched with the value segments from the segmentation grid. Together, these allies apply BPM technologies to link those components of the various databases that contain valuable customer information/knowledge in a manner that protects internal security. Now the parties involved have agreed to what knowledge will be made available and for what purposes, and they have established the means of access.

In the "distinctive" level, a sustainable competitive advantage is the objective. Here, the nucleus firm in the extended enterprise and its allies are offering differentiated products and services matched to the needs of the various segments in the grid. Through the sharing and analysis of mutually provided information, insight-driven interactions are a part of the scheme. The joint analysis of the data being transferred over the BPM-enabled extranet connecting the value chain constituents is providing knowledge unavailable to competing networks. As this model is considered, it is imperative that a firm which desires such an advanced position evaluate itself and determine where the organization and its network partners fall on the maturity scale. Then a determination can be made as to where the firm needs to be, and the business partners can begin building a plan to achieve that position.

RESPONDING TO THE CUSTOMER EXPERIENCE

The intelligent value chain that evolves will have many facets, but it will remain focused on customer satisfaction. The architecture that makes such a value chain possible is described in Figure 8.6. It progresses from the back-office systems, necessary to meet the needs of the customers, to the customer touch points so critical to the provision of value-added services. In between, a customer intelligence hub is at work, using BPM and providing the profile, rules management, events and treatments, and the quality data needed to enhance the ASCM/CRM systems.

New definitions are then brought to the benefits and values being delivered to the most strategic customers. Differentiated (often customized) answers to members of a particular segment's business problems are part of the delivery. Points of view are specific to each market segment. Solutions are comprised of a mix of tools, competencies, and offerings matched to actual needs. Specific solutions are packaged and delivered with a defined and quantifiable business value — measured across the entire value chain, and for the individual partners, using the economic value added tools described. The customer intelligence system at work synthesizes data consolidation and analytics so that a single view of the customer emerges, as well as individual customer analytics, which are used in profiling, evaluation, and modeling for success. A single up-to-date, integrated view of the customer relationship is constantly maintained, along with robust customer insights to tailor the correct treatment to the right customer at the right time.

There are three dimensions to customer intelligence, with specific features and advantages:

1. Customer information integration
   - Integration and rationalization of disparate customer data, to provide a persistent cross-channel data store to serve as a focal point for

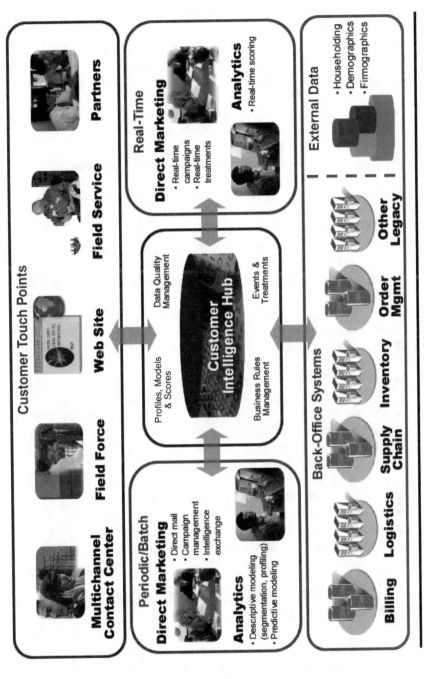

Figure 8.6. The Intelligent Value Chain Business Architecture

analytic processing and as a clearinghouse for multiple disparate touch points
- Establishment of relationships in the data to support analysis at the customer, prospect, household, and segment levels
- Development of an operations format for use of customer knowledge through all customer interaction points
- Development of event-based or delta-based sensing mechanisms to identify changes in front-end CRM systems, such as customer behavior or profile
- Transfer of information on event or delta to the hub-based repository for integration and consolidation
- Utilization of enterprise application integration or low-latency tools to move data from front-end systems to operational data storage

2. Customer insights: segmentation and modeling
- Ability to analyze cleansed and consolidated customer data to develop descriptive and/or predictive models
- Understanding of the economic or lifetime value of each individual customer
- Customer segmentation based on value, demographics, and behavioral information
- Quantification of each customer's responsiveness to marketing and other stimuli
- Identification of the appropriate treatment or offer for each customer, and delivery of this insight to front-end application
- Mining of vast amounts of data to identify hidden customer insights
- Capture and codification of analytical best practices in a business rules engine, to create intelligent recommendations in a near real-time environment

3. Customer insights: operationalization
- Ability to offer insights at the point of contact
- Products and services matched to individual customers
- Rules-driven customer interactions
- Differentiated service treatments for valuable customers

CONCLUSION: USING CUSTOMER INTELLIGENCE IS A KEY TO SUCCESS

Information abounds in most business organizations. The problem is that most of the important data is not used in an intelligent manner, because it is stored in nonintegrated databases and rarely shared across internal business units.

What must be done with this valuable information forms the basis of customer intelligence. That begins with a common definition of customers, a description of the tools to be applied, and the information integration architecture necessary to make the system a viable business enhancement process.

Knowledge is the key, and information about customers builds knowledge. Companies must manage their customer data better to be able to act upon customer knowledge. Figure 8.7 shows the customer intelligence maturity model, where we have arrayed the key operational improvement characteristics against the levels of maturity progression. From basic through foundational, core, and distinctive levels, the important areas where customer intelligence can bring a favorable impact are depicted, so a company can gauge how far along the continuum it wants to or should proceed. Firms need to assess where they are, what their competitors are doing, and then determine where they need to be in order to achieve the advantages that we have outlined.

	Basic	Foundational	Core	Distinctive
Customer Data Quality	• Manual data cleanup • Multiple customer IDs • Data not trusted by users	• Periodic batch process to clean up data • Common customer ID shared across systems	• Online search, lookup, validation rules applied at point of data entry • Cross-reference for IDs	• Real-time data quality maintenance using external vendor where appropriate
Customer Data Integration	• No 360 degree view of customer • Touch points acting independently • Customer aggregates and rollups performed manually	• Some point-to-point integration between front and back office • Some synchronization or data sharing across touch points	• Complete customer view available across all touch points with updates performed via periodic batch process	• Real-time, bi-directional integration between front and back office • Updates to single customer view available to all applications in real time
Analytic Sophistication	• Market segmentation only • No predictive analysis capabilities	• Known customer value • Segments defined by actual behavior • Treatments driven by segment	• Segment-specific predictive models • Tactical treatments driven by predictive outputs	• Products/services matched to interested customers • Differentiated service for valuable customers
Insight Operationalization	• Insights created by offline process • No integration to front-office applications • Offline direct marketing	• Insights created offline are manually fed into some front-office apps • No production process to update insights	• Insights available in most front-office apps • Model outputs (scores) updated through periodic batch process	• Insights updated with real-time scoring • Real-time, rules-driven, fully optimized customer interactions

Figure 8.7.   Customer Intelligence Maturity Model

9

USING PROCESS SIMULATION TO MINIMIZE THE RISK

Now that we have explained how to take a firm to the highest and most appropriate level of the supply chain maturity model, and how to track the savings from that model and the SCOR® model to financial statements, we are prepared to consider how to make sure the potential savings are real before engaging in what could be a high-risk transformation process. In this chapter, we will introduce the idea of using computer-based simulations to test business improvement changes before implementation and will suggest ways in which simulations can be used to help optimize processes and supply chains before encountering risk in the possible outcomes. Simulation will be explored and defined as a logical business implementation tool, with the inherent techniques explained along with numerous examples of how firms are getting the most benefit from the tool.

SIMULATION DEFINED

In essence, simulation provides a "virtual test bench" for process improvement, as it is a technique that focuses on quantitative measures that could result from various potential actions. It provides a representation of the process actions in a manner that is transparent to the business user, thereby generating a prediction of potential business performance — should the process, rules, and parameters be adopted or altered in practice. It provides a means of testing alternative solutions and outcomes before actually engaging in the introduction of the changed processing.

As we explore simulation, we will explain how it is used to gain value, its role within a business process management suite, and how it can be applied to business intelligence and management of the enterprise, through analysis of the resulting metrics. Additionally, we will show how optimization efforts can be applied with simulation to approach true process optimization in an automated manner. This level of optimization, implemented within business process management — enables the extended enterprise to more effectively manage its end-to-end supply chain and supporting business processes.

THE IMPORTANCE OF FOCUS ON PROCESS

To begin, managing businesses to achieve maximum value requires an understanding of what is meant by value. There has long been a quantitative approach to business management based on generally accepted accounting principles. As the various approaches have matured, they have become much more sophisticated, especially with the introduction of activity-based costing, balanced scorecards, and financial dashboards, all of which build on two fundamental concepts: when you can't measure it (however that might be), you can't improve it, and what you measure generally gets better. We now ask what it is that delivers goods or services with inherent value from a business or enterprise. It is a set of focused and coherent processes that bring satisfaction to the consumer. To improve a business, then, as you change the underlying processes, each process needs a set of measures which when folded together reflect attainment of the organization's goals as well as maximum customer satisfaction — the essential aims of an advanced supply chain system.

In the quest for such a condition, it has been said that there have been "five big ideas" in terms of operational management, notably:

- Introduction of the moving production line and standardized product by Henry Ford and Frederick Taylor
- Statistical control of quality by W. Edwards Deming
- Lean production by Toyota
- Theory of Constraints by Eli Goldratt
- Process focus by Michael Hammer and James Champy

Of these ideas, process focus is the only one that looks "end to end," while the others tend to work on single activities. In the current complicated business world, you cannot get far by focusing on a single task. Most equipment, for example, can be relied upon to deal with a single task effectively. On the other hand, process can be defined as "end-to-end work." It is an organized group of related tasks that work together to create value. All supply chain work fits this description and becomes process work (e.g., "order to cash" or new product introduction). The redesign of work on an end-to-end basis is central to improvement; it is the antidote to nonvalue-adding activities. Consequently, *process* can have the biggest impact on enterprise operations. Contemporary performance problems, as a result, are process problems, not task problems. We must use techniques suited to dealing with the process. This is where simulation comes in and offers the greatest value.

SIMULATION INTRODUCTION

The term simulation is used in a wide variety of applications. Weather modeling, business games, war gaming, and aircraft cockpit simulators can all be construed as forms of simulation. So what does the term simulation really mean to a business application? In a very general way, simulation can be defined as an imitation of a system. This is obviously a very general view and not particularly useful. Some simulation is enacted by human role play, while some involves physical models, and some is computer based. A key distinction is that the simulation of business processes involves the time dimension, which introduces *dynamic simulation.* An imitation on a computer of a system as it progresses through time represents one possible dynamic example, such as the building of an automobile.

The simulation technique introduced here is a discrete event simulation as opposed to any simulation model based purely on mathematical calculations. Mathematical calculations can be used as simulators or predictive devices and are the basis of many forecasting tools used in supply chain management. Discrete event simulation actually mimics the process in a very transparent way. As such, you can study how to achieve something rather than just get an indication of the outcome. This "how-to" feature is obviously of great benefit when determining the best way to manage the enterprise of which the firm is one constituent. Users of discrete event models can generally have confidence that when they modify the simulation of the process, the resulting outcome will be valid. It is not always obvious how to change a mathematical model to predict the effect of a change in how a business is operated.

Discrete event simulation can also be used to study systems designed to perform business or manufacturing processes, which can include specifically created physical systems (e.g., equipment or human activity or, even more

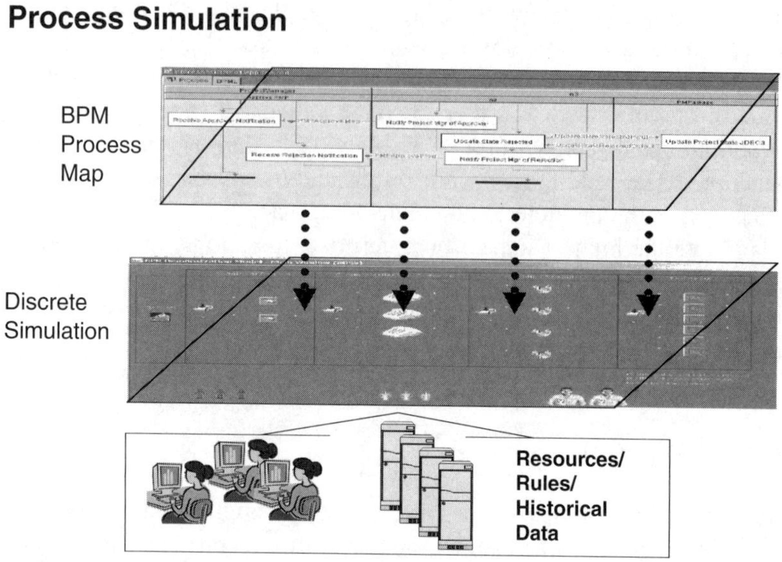

Process Simulation

BPM
Process
Map

Discrete
Simulation

Resources/
Rules/
Historical
Data

Figure 9.1. Business Process Map and Simulation Model

likely, a combination of people and equipment undertaking a set of defined tasks or a process). The relation of the business process to the simulation model is shown in Figure 9.1. For instance, service operations (banks, call centers, and supermarkets), manufacturing plants, distribution and transport systems, hospital emergency departments, and military operations all involve elements of both equipment and people. On some occasions, other effects need to be represented in the simulation models as well (e.g., weather affecting the progress of a ship). Normally, the effect, rather than the actual cause, is modeled or represented statistically.

Randomness is also an important factor. To move away from averages, real life is all about fluctuations in values, whether from market changes, traffic conditions, weather, and so forth. The time it takes a human to perform a task also varies, so it is important for any simulation to deal with randomness and statistical distributions. The purpose of a simulation model, moreover, affects how you build it, how you undertake the project, how you judge success, and how you know you have finished. Simulation projects might be about generating knowledge, investigating change strategies, developing control rules, etc. In each case, the model will be different because of the end use.

It is clear that different details will be required in a model to test whether option A is better than option B, as opposed to a user predicting to two decimal points the service level of option A. Simulation tends to involve the simplification of real systems into an appropriate model. A simulation simply predicts the performance of an operations system under a specific set of inputs. For instance, it might predict the average waiting time for telephone customers at a call center when a specific number of operators are employed. It is the job of the person using the simulation model to vary the inputs (the number of operators) and to run the model in order to determine the effect. As such, simulation is an experimental approach to modeling, which becomes a "what-if" analysis tool. The model user enters a scenario, and the model predicts the outcome. The model user continues to explore alternative scenarios until he or she has obtained sufficient understanding or identified how to improve the real system.

A computer simulation is a model that also mimics reality. Simulation involves the modeling of processes as they progress through time, according to the process definition, operating rules, constraints, and resources, often facilitated through computer processing of the data.

WHY SHOULD A FIRM SIMULATE POTENTIAL BUSINESS PROCESSES?

Simulation becomes valuable when there is variability of outcomes in the business processing or when parameters and rules are changed. Under these conditions, it is not easy to predict the outcomes, particularly when these factors change over time. For example, the famous "beer distribution game" represents a simple supply chain consisting of a retailer, wholesaler, and factory. The retailer orders cases of beer from the wholesaler, which in turn orders beer from the factory. There is a delay between placing an order and receiving the cases of beer. The game demonstrates that a small perturbation in the number of cases of beer sold by the retailer can cause large shifts in the quantity of cases stored and produced by the wholesaler and factory, respectively. Such a system is subject to dynamic complexity — like the majority of contemporary supply chains.

Of course, many operating systems are subject to both variability and dynamic complexity. Indeed, the variability of one component interacts with the variability of another to create dynamic complexity. In such situations, it is almost impossible to predict the overall performance of the system. Consider the simple example shown in Figure 9.2.

The level of customer service is simple to predict, since there is no variability in the system and because there is no interaction between components. The lack of interaction is a result of the service time being exactly the same for each customer, which means that there is no queuing or blocking. The

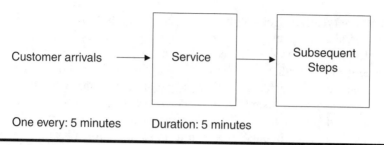

Customer arrivals   Service   Subsequent Steps

One every: 5 minutes   Duration: 5 minutes

Figure 9.2. Example of a System Subject to Variability

customers spend five minutes in the system; in fact, each customer spends exactly five minutes in the system.

Most activities, however, do not take exactly the same time every time. Assume that the times given above are averages, so customers arrive on average every five minutes and it takes on average five minutes to serve a customer. What is the average time a customer spends in the queue waiting to be served? This is not an easy question to answer, especially when subsequent steps are also considered, as there is variability in both customer arrivals and service times. Queues would develop between the steps and create consequential effects on performance. Most people, when asked for a specific time, tend to under-estimate the likely queuing time. Of course, the actual queuing time depends on many things even in the one-step system, such as:

- Variability in arrivals
- Variability in regular service time
- Variability in service time that might be affected by type of customer or time of day

Discrete event simulation provides the technique for evaluating such systems effectively, modeling the process from the "bottom up," starting at the required level of detail. Changing the process inherent in the model provides the ability to evaluate the effect of such a change. Operations that involve multiple processes and how they interact can also be modeled, by effectively linking multiple models together.

The benefits of simulation include:

- Risk reduction
- Greater understanding of process conditions and interactions
- Operating cost reduction
- Lead time reduction

- Faster plant/process changes
- Capital cost reduction
- Improved customer service

THE ADVANTAGES OF SIMULATION

The value to be gained with simulation begins as a step beyond experimentation with the real system. The simulation model removes the need to experiment with the real system. For instance, additional check-in desks could be placed in an airport departure area, or a change in the flow around a factory floor could be implemented. There are some obvious, and less obvious, reasons why simulation is preferable to such direct experimentation:

- **Cost** — Experimentation with the real system is likely to be very costly, as it is expensive to interrupt day-to-day operations in order to try out new ideas.
- **Time** — It is time consuming to experiment with a real system. It may take many weeks or months (possibly more) before a true reflection of the performance of the system can be obtained.
- **Control of the experimental conditions** — When comparing alternatives, it is useful to control the conditions under which the experiments are performed so that direct comparisons can be made. This is difficult when experimenting with the real system.
- **The real system does not exist** — A most obvious difficulty with real-world experimentation is that the real system may not yet exist or, more importantly, a modified system will engender better results.

Softer benefits include:

- **Fostering creativity** — Ideas which can produce considerable improvements often are never tried because of the time and cost constraints. The way models are traditionally presented and viewed results in a simplification of reality that can stifle creativity. With simulation, various ideas can be tried in an environment that is free of risk.
- **Creating knowledge and understanding** — Models help people to discuss issues, see effects, and marshal their collective thoughts.
- **Visualization and communication** — Visual simulations provide a powerful tool for enhancing communication and stimulating innovative ideas.
- **Consensus building** — Many simulation studies are performed in the light of differing opinions as to the way forward. Sitting opposing parties

around a simulation model of the problem situation can be a powerful means of sharing concerns and testing ideas with a view to obtaining a consensus of opinion.

GAINING VALUE FROM DISCRETE EVENT SIMULATION

Discrete event simulation can be used to model various processes, using entities or tokens moving through queues over time. Figure 9.3 describes the project steps in such an effort, beginning with establishment of the objectives and scope and proceeding to implementation. Typically, there are two types of queues involved in discrete event simulation. First, we have those where entities just wait, what we normally consider a queue. Second, we see those where entities reside while the "value-adding work" is carried out. These queues are often considered to be activities. The "event" occurs when the model changes state, normally instantaneously. This is typically the start or end of an activity or some form of interruption, such as a breakdown, lunch break, etc. At the end of an activity, the simulation would check to see what other activities can now begin as a result and schedule the end of that activity after allocating the required

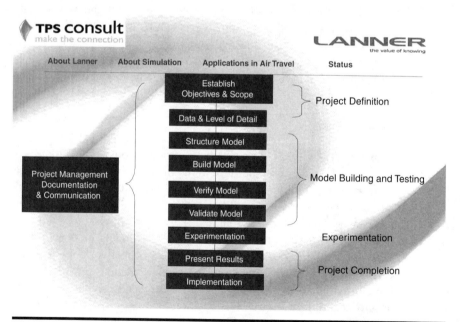

Figure 9.3. Project Steps

entities and other resources. The model then steps forward in time through a list of scheduled events, gathering statistics on performance and adding newly scheduled events.

CONSIDER THE ADVANTAGES OF EXPERIMENTATION

Consider a health service example, where the number of patients admitted to the clinic is a set number arriving according to a variable timing profile. To establish the weekly variability of utilizations of clinical staff, the model can be run for either a single run of 50 weeks *or* for 50 runs of a single week. Each of these runs will enable the variation of utilizations to be observed and confidence intervals calculated. Multiple runs are important. You would not judge the fairness of a coin by just one or two tosses!

It is true that the above model is simplistic, as indeed the profile of patients in such a model would probably vary in ways that would require more extensive modeling (e.g., the definition of a likely profile over a longer period to take into account current observed fluctuations from week to week). However, the experimentation options hold true: either a short period of experimentation can be repeated many times (with different random number sampling), or a model can be run for a much longer time and the range of random effects can be observed. How many repeats and how long to run a model are questions that can be answered by a combination of input and output data analysis. It is important when running a model that it experience all the randomness of the input data and that the output results "settle down" before conclusions are drawn.

At a second level, a user may wish to alter different parameters within a model to observe the different effects. Again, with this type of experimentation the range of results might also be important, once they are more accomplished through elongated or repeated experiments.

At a third level, a user may wish to compare one model with another. For example, in manufacturing there may be two investment options: production layout A and production layout B. These may indeed be the only options, although within each solution there may be parameter choices (e.g., containing buffer storage level options). In addition, there is again the optional value of establishing the potential range of results.

With all these levels of experimentation, there are a variety of experimental designs that can be used — different numbers of replications and different types of factorial experimental designs, ranging from full factorial to half factorial to Latin square-type designs where different parameter combinations are chosen. For different models, the model itself can simply be considered a different type of parameter.

CONSIDER OPTIMIZATION POSSIBILITIES

It is common within experimentation that the users want to find the best solution. But best can mean many different things — highest throughput levels, lowest costs, or highest services levels. Usually it means a specific combination of these types of factors. When searching for optimization, it is important to define what the best solution should look like. This is encapsulated in an objective function which can be as simple or as complex as required. This effort is not always a trivial exercise. Consider the problem of the optimization of a supply chain with many constituents, the extended enterprise model. Within a single enterprise, a balanced-scorecard-style approach is often used, balancing positives (savings) with negatives (costs). As the effort extends across multiple business constituents, the necessity is introduced to gather data in a more rigorous manner across several companies.

With the objective of the optimization established, the next step is to define what can vary (often called the optimization parameters). These parameters can be simple numbers, such as the number of staff employed on each shift, or can be flags indicating different complex control options or behaviors for a model. For example, it may be of interest to investigate flexible labor arrangements in a factory. For this type of situation, a flag can enact different resource rules in a model, one indicating separate defined resources and one indicating the use of any number of resource pools.

There may be constraints as well. For example, in a model that looks at an office business process, the staffing level for the seven different stages of a particular process may be between 20 and 30 people. However, the overall staff requirement must be below 190, due to space considerations. Other input data items, which need to be set for a simulation experiment, must be established for optimization, such as:

- How long should a model run be?
- Should there be a warm-up period for a model to allow it to reach equilibrium?
- How many replications of each set of parameters should be allowed?

Once the objective is defined and all the parameters and constraints have been set, then the model optimization is ready to run. An optimization is a series of experiments where the next experiment is determined based on the results found so far. The sequence can be determined and affected manually or automatically, using an optimization algorithm. Optimization can be thought of as intelligent experimentation, while optimization algorithms attempt to improve the results being achieved through learning the effects of adjusting each parameter in combination with other parameter settings. With a simulation model, an optimization algorithm typically works in the following way:

1. An initial set of parameter values is chosen, and one or more replication experiments is carried out with these values.
2. The results are obtained from the simulation runs, and the optimization algorithm then chooses another parameter set to try.
3. The new values are set and the next experiment set is run.
4. Steps 1 and 3 are repeated until either the algorithm is stopped manually or a set of defined finishing conditions is met.

With this type of heuristic algorithm, it is impossible to guarantee that the model will always find an optimal solution using a reduced experiment set. It is necessary, therefore, to define the simulation optimization solution space. The solution space of an experiment is often described as the multidimension plot of the objective function value against the parameter values. Thinking in simplistic terms with two parameters, it can be visualized as a terrain map, with the x and y coordinates as parameter values and the best result represented by the highest mountain or the deepest sea.

The most difficult type of solution space for any algorithm to search is one where the terrain is especially rugged, with many cliffs, chasms, crevasses, and all manner of quickly changing slopes and gradients. However, the majority of simulation models do not experience such wild variability. This is not to say that the landscape of a simulation solution space is bland, but usually there are trends apparent, rather than total randomness. Typically, optimization algorithms can learn quickly from the different sloping regions and can also learn about a relatively small number of complicated areas. Therefore, optimization algorithms can work well for typical simulation experiments.

The success of an optimization algorithm is often measured as the speed at which it can find an optimal or near-optimal solution. Speed is important because a simulation experiment is not an instantaneous calculation. Simulation experiments take time, and since it is not possible to conduct millions of experiments, a good algorithm must learn quickly in order to find good results in a reasonable time frame. Other useful and time-saving devices include:

- Simple analyze experiments to determine the variability of typical runs, which helps to determine the number of replications required
- An answer pool that saves data as the algorithm runs, enabling lightning-fast reruns and the reuse of results in other algorithmic runs

■ Tolerance settings to abort multiple runs if the first replication is so far removed from the optimal found as to make further replications a waste of time and unnecessary

■ Optional tracking for any other parameters that may be of interest in the results set from the simulation

EXAMPLES HELP ILLUSTRATE THE RESULTS

Figure 9.4 depicts a table of the first 38 experiments from a typical optimizer run, using Lanner's Witness Simulation package. Note the gradual improvement of the profitability function, which in this case is the objective function. The parameter settings are shown in the center columns and the tracked function values in italic on the right.

Figure 9.5 shows a view of the simulation run control and two of the reports to which it gives access. The value of the objective function found by each experiment is shown by the shaded line indicating the current optimal value discovered. The parameter values being used for the current experiment are shown on the right.

Figure 9.6 is a chart showing the variance of five different simulation replications. The nonregular shape of the polygon indicates that the simulation result varies between replications. Where different lines overlap, there is an indication that one parameter set is better than another — sometimes and sometimes not. Both of these results may illustrate options to effect better control on a process or indeed just reflect inherent variation that cannot be controlled.

Figure 9.7 is a table of confidence intervals for the result. These confidence intervals can also be obtained for any tracked function values. Figure 9.8 is a parameter analysis showing the average effect of each parameter value across all experiments measured in terms of improvement on the worst solution found. Such an analysis illustrates the sensitivity of different parameter settings.

DIFFERENT TYPES OF OPTIMIZATIONS

As outlined above, there are many different types of simulation experiments that can be conducted. Typically, optimization simulation is useful where there are ranges and sets of parameters and models that can be expressed as options and where the best solution can be expressed clearly. Model examples where optimization is used include:

	Evaluation	Profitability	Average Logging Time Value	TestandShip Time Value	StaffRulefor Logging Value	StaffRulefor Testing Value	StaffLevelA Quantity	StaffLevelB Quantity	Engineers Quantity	LoggingOrder Quantity	Checking Quantity	Testing Quantity	Engineer Utilization	ServiceLevel Percent	Profitability Service	Service Sigma Rating
1	0	5200	10	45	2	0	4	3	4		2	3	56.967	17.408	4172	0.557
2	1	5118	15	45	2	0	4	3	4		2	3	43.477	14.96	4090	0.493
3	2	4790	10	50	2	0	4	3	4		2	3	53.494	16.459	3774	0.52
4	3	7140	10	45	0	0	4	3	4		2	3	75.89	22.797	5680	0.703
5	4	6740	10	45	2	1	5	3	4		2	3	94.412	22.407	5536	0.738
6	5	5140	10	45	2	0	4	4	4		2	3	50.967	17.408	4112	0.557
7	6	7020	10	45	2	0	5	3	4		2	3	75.849	22.774	5768	0.748
8	7	5260	10	45	2	0	4	3	4		2	3	77.324	17.135	4232	0.547
9	8	5010	10	45	2	0	4	3	4		2	3	62.512	16.493	3994	0.522
10	9	5260	10	45	2	0	4	3	4		2	3	56.31	17.269	4220	0.55
11	10	5820	10	45	2	0	4	3	4		2	4	42.665	17.999	4698	0.579
12	11	7140	10	45	0	1	4	3	4		2	3	75.89	22.797	5680	0.703
13	12	7200	10	45	0	1	4	3	3		2	3	96.742	22.347	5506	0.737
14	13	7120	10	45	0	1	5	3	3		2	3	97.253	20.771	5690	0.678
15	14	7060	10	45	0	2	5	3	3		2	3	97.253	20.771	5706	0.672
16	15	7140	10	45	0	1	5	3	3		2	3	96.754	22.347	5876	0.737
17	16	7090	10	45	0	1	5	4	3		2	3	96.798	22.338	5808	0.736
18	17	7030	10	45	0	1	5	4	4		3	4	96.798	22.338	5766	0.736
19	18	7030	10	45	0	1	5	4	4		3	4	96.1	23.006	5778	0.76
20	19	9670	10	45	0	1	5	4	4		3	4	96.104	46.076	8502	1.4
21	20	9650	10	45	0	1	4	4	4		3	3	96.187	44.158	8450	1.391
22	21	9670	10	45	0	1	4	4	4		3	3	96.833	45.36	8494	1.381
23	22	8450	10	45	2	1	4	4	4		3	3	96.471	28.536	7086	0.931
24	23	7190	10	45	0	1	4	4	2		3	3	97.225	22.079	5910	0.724
25	24	9570	10	45	1	1	4	4	4		3	3	96.965	41.926	8330	1.295
26	25	9610	10	45	1	1	4	4	4		3	3	90.799	43.766	8406	1.341
27	26	9157	15	45	1	2	4	4	4		3	3	96.436	32.182	8394	1.004
28	27	7250	10	45	1	1	4	4	4		3	3	94.367	23.164	5666	0.764
29	28	9550	10	45	1	1	4	4	4		3	3	76.932	46.075	8394	1.399
30	29	8230	10	45	1	1	3	3	4		3	3	75.39	29.000	6002	0.946
31	30	9570	10	45	1	1	5	4	4		3	3	76.987	46.406	8414	1.41
32	31	6970	10	45	0	1	5	5	4		3	3	75.88	22.397	5702	0.733
33	32	7070	10	45	1	1	5	5	4		3	3	75.418	22.256	5786	0.729
34	33	9610	10	45	0	1	5	5	4		3	3	96.799	44.269	8380	1.354
35	34	9550	10	45	0	1	5	5	4		3	3	96.799	44.269	8380	1.354
36	35	9550	10	45	0	1	5	5	4		3	3	96.799	44.269	8380	1.394
37	36	10770	10	45	0	0	5	5	5		3	3	96.781	78.566	10214	2.353
38	37	10770	10	45	0	1	5	5	5		3	3	96.781	78.000	10214	2.353

Figure 9.4.   Experiments from a Typical Optimizer Run

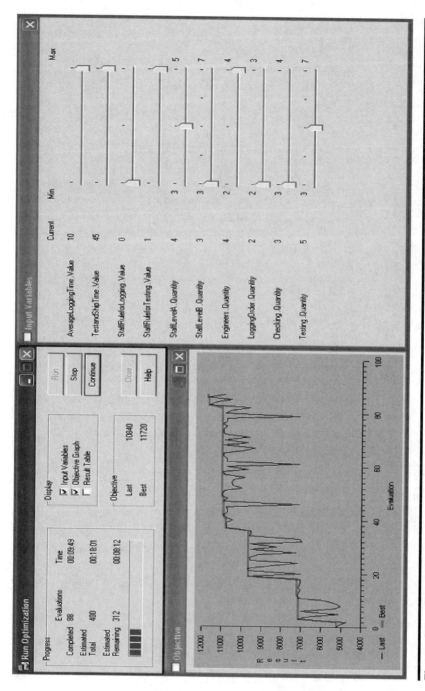

Figure 9.5. Simulation Run Control

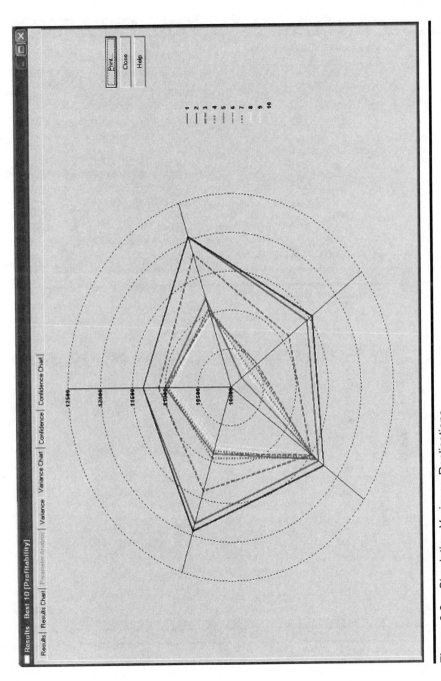

Figure 9.6. Simulation Variance Replications

Results - Best 500 [Profitability]

Results | Results Chart | Parameter Analysis | Variance | Variance Chart | Confidence | Confidence Chart

	Evaluation	Mean	90% Min	90% Max	95% Min	95% Max	99% Min	99% Max	
1	86	11720	11447.227	11992.773	11363.986	12076.014	11130.913	12309.087	
2	83	11660	11387.227	11932.773	11303.986	12016.014	11070.913	12249.087	
3	84	11380	11099.034	11660.966	11013.293	11746.707	10773.218	11986.782	
4	49	10997	10624.402	11369.598	10510.699	11483.301	10192.329	11801.671	
5	76	10980	10492.421	11467.579	10343.629	11616.371	9927.012	12032.988	
6	50	10977	10637.534	11316.466	10533.941	11420.059	10243.881	11710.119	
7	58	10960	10550.286	11369.714	10425.256	11494.744	10075.172	11844.828	
8	73	10940	10533.622	11346.378	10409.61	11470.39	10062.376	11817.624	
9	42	10930	10486.264	11373.736	10350.852	11509.148	9971.697	11888.303	
10	57	10898	10513.418	11282.582	10396.058	11399.942	10067.448	11728.552	
11	70	10880	10473.622	11286.378	10349.61	11410.39	10002.376	11757.624	
12	72	10880	10473.622	11286.378	10349.61	11410.39	10002.376	11757.624	
13	51	10877	10492.418	11261.582	10375.058	11378.942	10046.448	11707.552	
14	41	10860	10386.266	11313.734	10244.751	11455.249	9848.509	11851.491	

Sort Ascending
Sort Descending
Set Model
Set Suggested
Print..
Minitab
Close
Help

Figure 9.7. Confidence Intervals

Results - Best 500 [Profitability]

Results | Results Chart | Parameter Analysis | Variance | Variance Chart | Confidence | Confidence Chart

	Variable	Value	ObjFnAvg	% Benefit	
1	Testing.Quantity	7	10880	130.02	
2	Testing.Quantity	5	10744	127.14	
3	StaffLevelB.Quantity	5	10354	118.91	
4	Testing.Quantity	6	10058	112.65	
5	AverageLoggingTime.Value	15	9889	109.07	
6	Engineers.Quantity	3	9787	106.91	
7	StaffRuleforLogging.Value	0	9625	103.49	
8	StaffLevelA.Quantity	3	9505	100.96	
9	Checking.Quantity	4	9351	97.70	
10	LoggingOrder.Quantity	3	9230	95.14	
11	StaffRuleforLogging.Value	1	9226	95.04	
12	StaffRuleforTesting.Value	1	9196	94.42	
13	TestandShipTime.Value	50	9025	90.79	
14	StaffLevelA.Quantity	4	8982	89.89	
15	TestandShipTime.Value	45	8956	89.34	
16	Checking.Quantity	3	8912	88.42	
17	StaffRuleforTesting.Value	0	8789	85.82	
18	AverageLoggingTime.Value	10	8780	85.62	
19	Testing.Quantity	4	8775	85.51	
20	StaffLevelA.Quantity	5	8758	85.15	
21	LoggingOrder.Quantity	2	8717	84.29	
22	StaffLevelB.Quantity	4	8661	83.12	
23	Engineers.Quantity	4	7528	59.16	
24	Engineers.Quantity	2	7113	50.39	
25	StaffRuleforLogging.Value	2	6662	40.85	
26	StaffLevelB.Quantity	3	6408	35.47	
27	Testing.Quantity	3	6337	33.97	
28					

Sort Ascending
Sort Descending
Print..
Minitab
Close
Help

Figure 9.8. Parameter Analysis

- Improving key performance indicators, for example, which means organizational control and process flows can be changed to improve performance

  □ Determining resource numbers, such as how many people should be employed, with what skills, and on what shift patterns

  □ Determining storage requirements, to answer what buffers should be present at each stage of the production process for a typical schedule

  □ Determining control and priorities, for example, in a hospital to deal with patients in the most efficient and correct order

  □ Determining layout, to ascertain what amount, for example, should be placed in chemical tanks for the production process to maximize the effectiveness of the subsequent transfer for usage

  □ Determining options for improvement, such as establishing in a Six Sigma project which machine cycle time it would make most sense to improve in order to maximize jobs per hour throughput for the whole line

SIMULATION AND KEY PERFORMANCE INDICATORS

Key performance indicators (KPIs) help an organization define and measure progress toward organizational goals. Once an organization has analyzed its mission, identified all its stakeholders, and defined its goals, it needs a way to measure progress toward those goals. KPIs are typically used for that purpose as measurements that are quantifiable, agreed to beforehand, and reflect the critical success factors of an organization. They will differ depending on the organization, but within a vertical industry sector some should be common and used to benchmark a company's performance against competitors.

A business needs to set targets for each KPI. A company goal to be the employer of choice might include a KPI of "people turnover rate." After the KPI has been defined as "the number of voluntary resignations and terminations for poor performance, divided by the total number of employees at the beginning of the period" and a way to measure it has been set up by collecting the information in the human resources system, the target has to be established. "Reduce people turnover by 5% per year" is a clear target that everyone will understand and be able to take specific action to accomplish. The question then becomes: What are the specific actions that will deliver the objective?

Business process management is about managing corporate or business performance through managing those processes which drive the firm to the desired overall objectives. KPIs therefore need to be designed at various levels in order to be relevant at an individual process level. While the KPI defined above concerning turnover rate might be relevant for the board, it is not appropriate for the supervisor of the human resources department's contact center. That center's performance may well have an effect on turnover rate, but it requires its own process-level KPIs, perhaps based on responding successfully to employee questions, in order for it to be managed effectively.

Simulation and optimization supporting the identification and design of KPIs have been used extensively to help improve processes and consequently to ultimately improve business performance. Simulation can be used to both help define and set targets for a KPI. This feature is in addition to the use of simulation to identify process changes (resources, rules, and structural needs), which deliver improved performance. Simulation aids the definition of a KPI (that is, the equation and data of which it is composed), through examining the reaction of that KPI under different circumstances. This procedure ensures that the KPI does in fact properly connect to the organization's goals and will help drive the appropriate behavior and reaction to different potential events. Most KPIs are developed directly from the overall goals or critical success factors of the organization. Process-based KPIs tend to be focused on results or outputs from a process, such as customers served per hour or claims processed.

Simulation is also valuable in the setting of targets against KPIs (such as the number of calls to be answered per hour), because the simulation can accurately assess what is achievable in theory by the process and resources employed. Many KPIs are now presented as part of a "digital cockpit," using gauge or dashboard-style displays. Depending on the KPI and the targets selected, they often include some use of zones to highlight acceptable levels (e.g., green, yellow, and red). It is important when calibrating these gauges to understand the effect of natural fluctuations due to inherent randomness within processes or behavior. Simulation can support this calibration to help ensure that unnecessary reaction to natural short-term fluctuations does not occur.

Processes have a series of steps, and at each step measurement can be taken to better understand the behavior of the process under different circumstances. Identification of key points in the process and the relevance of specific measures (probes) in predicting failure of the process can provide significant opportunity for improved process management and business value. Effective process monitoring through probes at specific points in the process involves watching factors to give prior warning that the target levels are under threat. Simulation is able to help identify these probes and the threshold values that indicate a deterioration of the process which will ultimately end in an unacceptable KPI.

SUMMARY

Management through a process focus and by using validated metrics is a valuable and proven methodology to help drive a successful enterprise. Through their natural processing and quantitative focus, simulation and optimization provide a natural enabling tool to help connect the top-level balanced scorecard and corporate performance management approaches to detailed business process analysis contained within the business process management umbrella. They also pave the way for supporting continuous process improvement as more and more companies become properly process enabled.

10

STEPS FORWARD

In this concluding chapter, we want to revisit some of the main themes and draw conclusions that lead to a productive go-forward effort. First, it is important to understand that a supply chain effort must be focused on four key business drivers:

- Reducing costs as low as possible without compromising the ability to satisfy customers
- Making optimum use of assets, whether they be owned by the firm or one of its important business allies
- Using the benefits of the supply chain improvements to satisfy customers and consumers
- Increasing revenues with those customers most beneficial to the firm and its business allies

Second, to achieve optimum results across those needs, a firm must follow some form of framework or model and get its internal house in order and then progress, with the help of willing and trusted external business allies, toward a predetermined position of excellence, which significantly improves the metrics used to track progress. Third, the internal and external progress should be facilitated with the use of business process management and business process management systems that transfer important knowledge quickly and easily between the constituents in what becomes an intelligent value network. This network should be focused on having distinguishing characteristics in the eyes of the most important customers and consumer groups. Fourth, with the preceding steps accomplished and better metrics achieved, the actual benefits should be clearly visible in improving the profit and loss statements and balance sheets of all the partners in the value chain.

Those are the fundamentals of a supply chain management and advanced supply chain management effort. We have outlined how a firm can use the supply chain maturity model as a framework to guide its effort and to gauge progress. The SCOR® model, with enhancements, can be equally helpful in finding the route to better performance and near optimized operating conditions. Simulation can be applied to minimize the risk before making actual changes or to test various scenarios. When accomplishments are recorded, however, there must be a means to document the improvements on the financial statements. That is the essence of a successful effort.

THE GO-FORWARD STEPS

With an appreciation for the views expressed in this book, and the advice given for making progress and finding the values in the financial statements, we are prepared to conclude with the steps that have been mentioned before but which are worthy of repeating as they are appropriate to consider as a means of moving forward. These steps begin with some form of an *awareness* session, with attendees representing the key decision makers in the organization. Most supply chain efforts fail to reap the full benefits because important executives are not aware of what the results can mean to their areas of the business and therefore withhold full support. An awareness session focused on explaining the basic concepts, presenting the roadmap to be followed, and showing at least order-of-magnitude metrics covering the impact to the financial statements and to the individual executives' areas of responsibility is an imperative action.

Next, there should be an effort led by the CEO and the designated supply chain leader to develop *alignment* around the vision, strategy, and action plan, which is developed to guide the effort. Once again, we have found from experience that there is too much cosmetic endorsement of supply chain efforts, rather than true alignment and support for what can be accomplished. Going forward without some solid assurance that alignment has been achieved across the senior management team is a formula for failure.

With alignment, an *action* plan can be developed and implementation started. The correct actions seem to follow the earlier steps, as awareness and alignment tend to assure the effort is coordinated, supported, and directed, crucial ingredients for success.

As most companies have some form of effort under way, the next important step to guide the action plan is *calibration,* that is, finding out where the business stands versus where it could be under near optimized conditions. Then the value to the business of closing the gap and exceeding the most important metrics can be determined. A dashboard displaying the desired future metrics can also be created to track the improved performance.

The calibration exercise can be extremely deep and complex, using consultants to help in the search for meaningful benchmarks against industry competitors or industry leaders in an area of particular importance. On the other hand, it can be conducted at a high level with a very simple matrix, like that shown in Figure 10.1. Here the firm is asked to assemble an audience of knowledgeable people representing the various business functions and business units who then rate the firm's position on the matrix.

The session begins with some questions: How do we view the current and future industry and market conditions? Where do we currently stand with our improvement efforts? What does the future hold for us, and how will we take advantage of our supply chain efforts to provide value to our customers and end consumers? The chart is then presented as a means of determining where the firm or business unit stands versus where it could be in the future in the various functional areas. Each column represents where there could be movement to a higher level, if it is appropriate to achieve that position. The chart is read from left to right, with the right-most column representing the future, or transformed company position. Some companies will have activities in multiple columns or business units with varying positions. The idea is to reach some form of majority acceptance of current versus "could-be" level. To add a scoring technique, accept the consensus position and assign a value as follows:

- 1 point for each cell in column 1
- 3 points for each cell in column 2
- 5 points for each cell in column 3
- 7 points for each cell in column 4

The maximum scoring is 49, unless you decide to add other functions or categories.

RULE OF THUMB SCORING

To gauge the firm's overall position, a simple scoring mechanism can be used. We have had success applying the following chart and have added some general comments on the scores achieved:

7–15   Internally focused. You are lagging in the market. Look into advanced supply chain management tools.

Follow the flow and determine: Where is your company? Or where is your business unit? Go through the matrix and put a check in the section that best describes your company's position. As a rule of thumb, if you are undecided about which cell to choose, choose the cell on the left.

		Level 1 & 2	Level 3	Level 4	Level 5	Notes
	Business Application	Internal Optimization/ Supply Chain Optimization	Advanced Supply Chain Management	Value Network	Total Network Connectivity	
1	Purchasing, Procurement and Sourcing					
2	Logistics, Transportation and Warehousing					
3	Forecasting, Planning, Scheduling, Inventory and Manufacturing					
4	Marketing, Sales, & Customer Service					
5	Information Technology					
6	Design and Development of Products and/or Services					
7	Supplier/ Customer Collaboration — SRM/CRM					
8	Human Resources					
	Column Totals					Total
	Score	1	3	5	7	

Figure 10.1. Calibration Chart

16–30   Some external focus. Your company is with the majority of firms. You are likely well positioned to move quickly ahead. As an example, consider the possibility of providing vendor-managed inventory/scheduling via a secure extranet.

31–40   Good external focus. You are well positioned as a potential market leader. Examine the possibility of taking the next step of forming a value chain with key customers and suppliers in order to focus on and fulfill the needs of a particular market segment.

41–49   Congratulations! You are a market leader. Look into your one or two areas of improvement.

More complex techniques can be used, with much greater detail contained in the questioning and scoring, but we find that they lead to the same general conclusions.

The final step in calibration is to make an initial assessment of the value of moving from the firm's current state to the desired future state. The figures in earlier chapters will give some indication of the potential, and the reader will be able to augment these values with his or her own experiences.

EXAMPLE FROM THE GLOBAL AUTOMOTIVE INDUSTRY

There is one more principle that helps when doing a calibration. Seeing a future state is not helpful unless it leads to three things:

- Understanding how the future state will impact the business
- Knowing where the firm is in relationship to the future
- Moving toward that future in the most effective manner

The automotive industry can be used as an example of an industry trying to transform itself while satisfying the three requirements. Several manufacturers, including General Motors, Ford, and Toyota, for example, are all moving toward full network connectivity as aggressively as they can. Knowing that part of the future belongs to the firm that can respond to the consumer in the shortest interval with what each consumer wants, these companies are hard at work improving their supply chains to gain competence in that direction. The savings of many days of inventory combined with meeting the demands of car buyers and dealers in a shorter cycle time are ideal level 4 and 5 objectives.

The combination of the efficiencies of adding first- and second-tier suppliers to the value network and linking dealers and customers electronically is compelling. The automobile industry is one of the industries attempting a transformation along the guidelines provided by the maturity model. The participants are faced, however, with many barriers as they look to their position versus the expected future needs: their own internal processes, investment in capital goods and technology, regulatory issues, extraordinarily diverse markets, global overcapacity, changing consumer expectations, and more.

Even with all those barriers, key executives realize they must change the way they do business if they are to move effectively into the future. For an auto company to become the driver in a level 5 network means adding a build-to-order capability, which would allow a consumer to customize a vehicle order, receive a firm price and delivery commitment, and reduce the time to deliver a vehicle to days. Build to order is a change from the push model. The push model is what was experienced: fill the distribution channel with vehicles, then influence buyers to make a purchase using extensive ad campaigns, rebates, low-cost financing, and more. An ideal pull model means the consumer will order a vehicle and it will be produced after the order is placed. The consumer gets exactly what he or she wants, and the manufacturer and the balance of the supply chain constituents are assured of a sale. A simple mechanism like the one shown in Figure 10.1 can be very useful in establishing the areas needing improvement and in setting up the improvement sequence for a similar effort, especially when conducted in a cross-enterprise manner.

IDENTIFY THE GAPS FROM CURRENT TO DESIRED FUTURE STATE

With some form of fundamental calibration completed, the next step is to flesh out more specific details on the firm's positioning against its industry competitors, so the leaders can chart a meaningful route forward from the identified current state to the desired future state. That exercise begins with a look at where the firm stands vis-à-vis the competition and industry leaders. We prefer to start with an analysis such as that performed by Stratascope, a Marlborough, Massachusetts–based firm specializing in data-based research and selling tools. This firm can provide a profile compiled from its five-year histories of 25,000 companies, created using publicly available quarterly financial metrics, as filed with the Securities and Exchange Commission.

To illustrate our suggestion, we will use material provided by CEO Bruce Brien on five companies providing products to the automotive original equipment manufacturers and aftermarket car parts suppliers. The companies studied were Borg Warner, Bosch Automotive Systems Corporation, Delphi Corporation, Johnson Controls, and Tenneco Automotive. Figure 10.2 is an overview of the five firms and the competitive landscape that existed at the quarters indicated. We begin by noting a significant range in return on capital investment, an immediate opportunity area for one or more firms. The net operating profit against capital invested (NOPACI) also appears to be low by other industry standards.

	TENNECO AUTO	BORG WARNER	BOSCH AUTOMA	DELPHI CORP	JOHNSON CONT
City	Lake Forest	Chicago	TOKYO 150-8360	Troy	Milwaukee
State	IL	IL		MI	WI
Country	USA	USA	JPN	USA	USA
Ticker	TEN	BWA	ZEX	DPH	JCI
Source	COREDATA	COREDATA	COREDATA	COREDATA	COREDATA
Quarter	2004Q1	2004Q1	2003Q4	2004Q2	2004Q2
Fiscal Year End Month	12	12	3	12	9
Beta	1.80	1.00	1.00	1.00	0.70
Employees	19,139	14,300	3,589	190,000	118,000
SIC Code	3714	3714	3714	3714	3714
Quartile	2	2	1	4	2
Currency	USD	USD	JPY	USD	USD
Revenue	3,879	3,197	334,710	28,780	24,964
NOPAT Percent	2.2%	5.5%	5.8%	0.1%	3.1%
Capital Turnover	2.02	1.24	2.10	1.86	2.83
Return on Capital Invested	4.6%	6.8%	12.2%	0.2%	8.8%
NOPACI Percent	1%	0%	3%	-2%	1%
EPS	0.59	3.28	0.00	(0.15)	3.87
Revenue per Employee	202,675	223,538	93,259,961	151,474	211,558
Oper Income per Employee	9,039	18,944	6,081,638	47	9,710
Net Income per Employee	1,254	12,713	5,624,965	(453)	6,208

Figure 10.2. Automotive Parts Competitive Landscape: Comments from the Annual Report (Source: Stratascope)

It is generally useful to also extract statements from the company's annual report to show how supply chain can help meet objectives and improve some of the metrics displayed. Most importantly, we typically use this type of information to point out the degree to which supply chain improvements, based on actual data, can increase the earnings per share.

Drawing out one of the players in this analysis, we can see in Figure 10.3 that Tenneco Automotive is exhibiting performance in several areas that could warrant further discussion. Similar conditions could be found for the other four competitors. The areas of importance are noted in the circles, with the size indicating the relative amount of importance. They include revenue growth; selling, general, and administrative expenses; inventory levels; and fixed asset utilization. All of these areas were found to be in alignment with management's discussion in the firm's annual report.

As we take a look at revenue growth, we see in Figure 10.4 that the firms are managing growth and approach the market differently. This observation is not an indictment, but could point to ways to improve desired performance. One immediate consideration is how other competitors are managing their growth

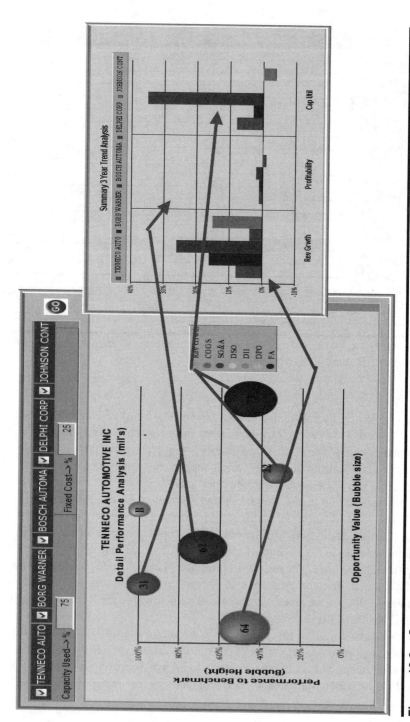

Figure 10.3. Summary Profile

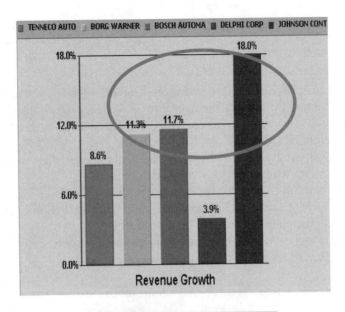

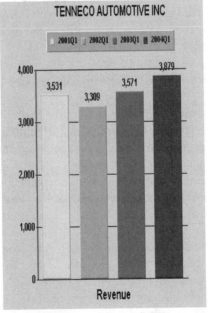

Figure 10.4. Revenue Growth Discussion

more aggressively than Delphi Corporation and how Johnson Controls has become the current leader. Supply chain suggestions that might be appropriate include:

- Overcoming difficulties in introducing or offering new products to customers by making them visible and readily accessible; providing easy access to capable-to-promise capabilities and available-to-promise inventories
- Reducing out-of-stocks by using the visibility to match supplies from inventory with actual current needs
- Cross-selling different brands through different channels and banners
- Assuring customer-scheduled order ship dates will be met with a very high percentage with high order fill ratios
- Handling sales quotations and order processing electronically in an error-free environment and industry best cycle times

In the area of selling, general, and administrative expenses, we see further opportunity, as illustrated in Figure 10.5. Operating expenses are apparently being managed differently by these companies, with Delphi setting the competitive pace, followed by Johnson Controls. Some of the issues meriting discussion could include:

- Improving the ability to tie the actual costs to the behaviors that account for them
- Reducing overall spend by increasing the visibility into the aggregated spending in a meaningful manner that highlights the opportunities
- Shortening the new product development and introduction cycle with increased predictability
- Reducing selling expenses with greater customer intelligence and customer relationship management matched to most likely improvement areas

Progressing into a discussion of inventory levels, we can see in Figure 10.6 that there is another wide variance, indicating some firms are better able to reduce the days of stock in inventory. Johnson Controls stands out in this category. With inventory an issue of such importance, aggressive management is an imperative. Issues worth pursuing could include:

- Using data collection and analysis from supply chain activities that are out of sync with the physical activities driving them, such as inventory imbalances and transaction data that are not trusted

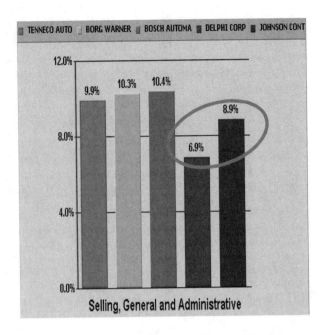

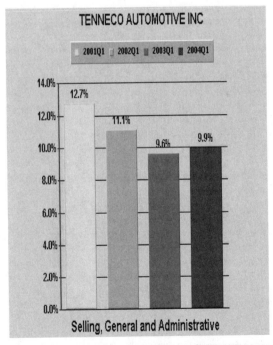

Figure 10.5.   Selling, General, and Administrative Expenses Discussion

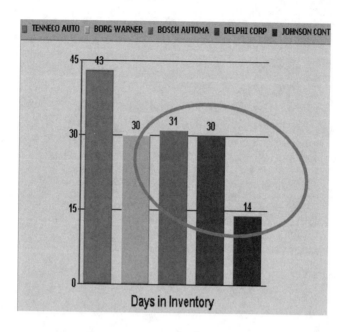

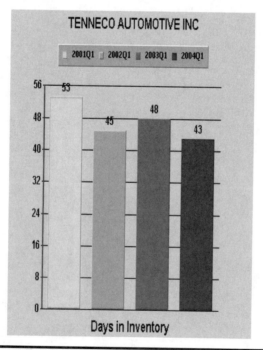

Figure 10.6.   Inventory Levels Discussion

- Managing the balance between cost and service levels in a dynamic environment to assure the service matches the customer need and willingness to pay for extra costs
- Configuring the level of shared information between trading partners to achieve better matching of demand and supply; bringing key customers into sales and operations planning and advanced planning and scheduling meetings
- Improving forecast accuracy and reducing variance on core products through knowledge sharing across customer-supplier databases

Finally, we turn to the fixed asset utilization part of the discussion. Figure 10.7 shows Johnson Controls and Delphi in the lead here. They appear to be investing fewer assets to deliver a dollar of revenue. Now the discussion points could cover:

- Is detailed shop floor schedule sequencing being done manually or in a batch mode, limiting dynamic resequencing based on the most current information?
- Are there too many trips to the central warehouse for small quantities?
- Is there an inability to optimize the location of manufacturing between plants, contract manufacturers, and other suppliers?
- Has the utilization of assets been suboptimized across the value chain network?

With this type of discussion, the management team could move to a discovery session to explore the root causes of the value gaps that have been identified. This session usually leads to identification of alternative solution scenarios for each of the highest priority issues. Action teams can then be assigned for implementation of the appropriate solutions.

DEEPER ANALYSIS TOOLS ARE AVAILABLE

If the firm is not averse to using external help in the calibration and analysis phase, there are a variety of tools available. CSC offers a diagnostic termed high-impact supply chain assessment, which includes a facilitated awareness session with benchmarking against competitors and industry leaders. Figure 10.8 is a sample of the tools used in that assessment. As the calibration is conducted in a specific area, such as procurement and strategic sourcing, capabilities are discussed for each level of the maturity model. Other charts are supplied for the other functions of importance to the analyzing company.

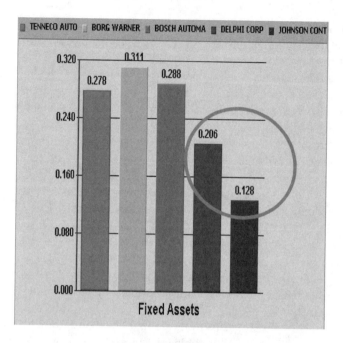

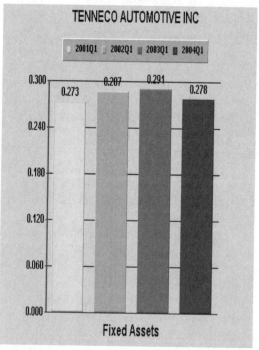

Figure 10.7. Fixed Asset Utilization Discussion

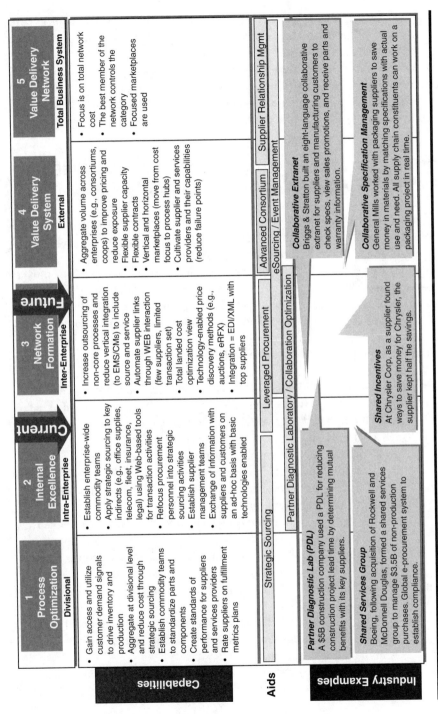

Figure 10.8. Procurement and Strategic Sourcing

The terms and examples used in the sourcing chart can be modified for industry-special needs. As the levels progress, available assistance is indicated by the bar in the middle of the diagram. Examples of what leading companies have done are also included in the bottom industry example section. This tool has proven to be very valuable as a firm determines its current and future desired state. From this simple diagnostic, most companies are able to begin planning an improvement process that takes the firm to that future state.

BUSINESS PROCESS MANAGEMENT SYSTEMS RELEASE THE BENEFITS OF ADVANCED SUPPLY CHAIN MANAGEMENT

With an understanding of the firm's position on the maturity model versus where it needs to be to catch the leaders or to gain a competitive advantage, the tool of most importance in supplying the knowledge necessary to enhance the targeted processing is business process management. As the firm moves, with the help of carefully selected business allies, toward optimized conditions, business process management will provide the means of accessing the databases across the whole value chain to extract vital information from components of the partnering firms' information. There are three key factors inherent in this business process management systems activity:

- **Collaboration** — Between internal colleagues, departments, and business units and between external constituents across the extended enterprise. Joint objectives are established through this collaboration to attain such goals as:
  - ☐ Realize a 35% return on investment on a collaborative planning, forecasting, and replenishment effort
  - ☐ Analyze real-time business information to reduce cycle times in key product development and delivery efforts
  - ☐ Link the work flow processing so there is full visibility into the flow of inventories and finished products
  - ☐ Connect the available-to-promise and capable-to-promise inventories with trading partners' and key customers' demand signals
- **Automation** — Of all important business processes, so the business allies can:
  - ☐ Eliminate slow and error-prone manual systems and model, deploy, and manage automated processes of importance to the network and its intended customers
  - ☐ Map legacy business objects so the process flows can be handled in a more efficient manner

- **Integration** — To drive up the values received from existing information technology investments to the operating units needing supply chain improvement, through:
  - ☐ Connection of the databases, legacy systems, and enterprise applications of importance
  - ☐ Work flows conducted with a single, unified view of the legacy business objects
  - ☐ Secure access to back-office systems

CONCLUSIONS

The benefits are only limited by the imagination of those seeking the improved state. As we have mentioned, the results of the joint CSC/*Supply Chain Management Review* survey have confirmed the possible improvements to costs and revenues. That survey also concluded that there is significant room for further improvement. Key process improvements cited by a large number of firms that have indicated achievement of benefits from their efforts include:

- Reduction in elapsed time for key process steps (order to cash, new product introduction, and so forth)
- Higher productivity for the manufacturing process and individual employees
- Improved quality and reduction of errors and need for reconciliation
- Reduction or elimination of process steps
- Greater customer satisfaction
- Automation of routine administrative tasks
- Reduced cost per transaction
- Reduced shrinkage and waste
- Reduced risk in attempting new process changes

The approach we have presented here is based not on hearsay, but on firsthand knowledge of what can work — for firms of any size in any industry. It requires a guiding framework, willing business allies, a trusting atmosphere, and business process management systems as the catalyst. The prize is a truly value-managed enterprise.

BIBLIOGRAPHY

Aberdeen Group, Collaborative Planning: You Can't Optimize What You Can't See, White Paper, November 13, 2002.

Acohido, Byron, Boeing outsources 7E7 work to Japanese companies, *USA Today*, November 21, 2003, p. 2B.

Andrew, James P. and Sirkin, Harold, Innovating for cash, *Harvard Business Review*, September 2003, pp. 76–83.

Asgekar, Vinay, The demand driven supply network, *Supply Chain Management Review*, April 2004, pp. 15–16.

Bacheldor, Beth, Keep supply chains flowing, *Information Week*, April 14, 2003, p. 57.

Bernhard, Frank and Vittori, Wendy, The straight path to profitability, *Supply Chain Management Review*, July/August 2004, pp. 58–64.

Clayton, Raymond A., *Purchasing: Pipeline to Corporate Profits*, Clayton Advisory Management Practice, reported in *Supply Chain Management Review*, November/December 2002, p. 35.

Edmundson, Gail, Customization — BMW, *Business Week*, November 24, 2003, p. 94.

Fontanella, John, Asgekar, Vinay, and Swanton, Bill, Demand-driven supply network: Striving for supply chain transparency, *AMR Research Report*, January 22, 2004, pp. 1–8.

Forrester White Paper, Supply Chain Management Processes Replace Apps: 2003 to 2008, Forrester Research, Cambridge, MA, December 2002.

Greenfield, Avi, Shin, Dennis, and O'Connor, Christine, Good management takes planning, *Information Week*, June 21, 2004, pp. 65–69.

Gutmann, Kirk, How GM is accelerating vehicle development, *Supply Chain Management Review*, May/June 2003, pp. 34–39.

Harris, Richard, *Business Process Management: Maximize Competitive Advantage by Aligning BPM with Your Business Objectives*, Gartner, Sydney, 2004.

Kontzer, Tony, Learning to share, *Information Week*, May 5, 2003, pp. 29–37.

Koudal, Peter and Lavien, Todd, Profits in the balance, *Optimize.com,* June 2003, pp. 81–89.

Poirier, Charles C., *Advanced Supply Chain Management,* Berrett-Koehler, San Francisco, 1999.

Poirier, Charles C., Achieving supply chain connectivity, *Supply Chain Management Review,* November/December 2002, pp. 16–22.

Poirier, Charles C., *Using Models to Improve the Supply Chain,* St. Lucie Press, Boca Raton, FL, 2004.

Poirier, Charles C. and Bauer, Michael J., *E-Supply Chain,* Berrett Koehler, San Francisco, 2000.

Poirier, Charles C., Ferrara, Lynette, Hayden, Francis, and Neal, Douglas, *The Networked Supply Chain,* J. Ross Publishing, Boca Raton, FL, 2004.

Quinn, Frank, How are we doing? A survey of supply chain progress, *Supply Chain Management Review,* November/December 2004, pp. 24–31.

Rudzki, Robert, The advantages of partnering, *Supply Chain Management Review,* March 2004, pp. 44–51.

Sullivan, Laurie, Ready to roll, *Information Week,* March 8, 2004, pp. 45–48.

Treacy, Michael and Wiersema, Fred, *Disciplines of the Market Leaders,* Addison-Wesley, Reading, MA, 1995.

INDEX

A

Aboulafia, Richard, 104
Action plan, 210–211
Activity-based costing, 40
Advanced business performance, 18–19
Advanced planning and scheduling (APS), 6, 19
Advanced supply chain management (ASCM)
  benefits of, 224–225
  coordinated plan, 121–122
  customer relationship management and, 168, 176, 178, 179, 181, 183, 184
  enabling technology, 121–122
  end line for, 12
  in level 4, 6–7
  potential for, 56–57
  survey results, 9
Agility, 51
Airbus, 96–97
Alenia Aircraft, 104
Algorithms, 198–199
Allders Department Store, 122
Alltel, 123
AMR Research, 35, 158, 166
Analysis tools, 221, 223
Applications, 114

APS. See Advanced planning and scheduling
Architectures, 150, 185
ArvinMeritor™
  BPM in supply chain, 139, 141
  business process automation, 145
  collaboration principles, 133, 138
  consolidation and transformation, 139, 142, 143, 144, 145, 146, 147, 148
  digital engineering, 147
  document management, 143
  governance model, 139, 140
  internal process validation, 139, 141
  knowledge management, 148
  SCOR® model and, 139
  team project collaboration, 144
ASCM. See Advanced supply chain management
Asset reclamation options, 170
Asset utilization, 159
Auctions, 108
Augmented SCOR®, 25–31. See also SCOR® model
Automated process management systems, 141. See also Business process management systems
Automotive industry, 97, 133, 138, 213–214
Autonomous horizontal businesses, 65–66

Available capacity, 94
Avnet, 123

B

B2B portals, 108
B2C portals, 108
Balanced scorecard, 40, 198, 207
Balance sheet
  actions taken and, 63
  examples, 56–57
  process-based, 67, 70
  RONA and, 68
Barclays, 104
Batched demand, 94
Batched requirements, 92–93
Batch sizes, 98
Bauer, Michael, 128, 131, 133, 138
"Beer distribution game," 193
Benchmarking, 18, 205, 211
Best Buy, 122
Best-in-class "silo" processes, 136
Best practices, 89, 145
BizTalk, 145
Black, Alex, 174
BMW, 103–104
Boeing, 37, 96, 104, 156, 158
Borg Warner, 214
Bosch Automotive Systems Corp., 214
BPM. *See* Business process management
BPMI, 141
BPML. *See* Business Process Management
  Languages
BPMS. *See* Business process management
  systems
Brien, Bruce, 214
British Aerospace, 98
Browser-based interfaces, 152
Buffer stocks, 62
Build-to-order products, 107, 214
"Build-to-order" system, 103
Business
  agility, 107
  architecture, 185
  intelligence, 159
  partners, 156, 166, 167, 171
  process challenges, 62

process map, 192
process model, generic, 12–17
strategy, 158
Business allies
  careful selection of, 101, 105, 106, 120,
    224
  example companies, 102, 104
  external, 156
  in manufacturing or processing area, 62
  trusted, 167, 175
  willing, 173
Business process management (BPM)
  across end-to-end processing, 178
  as breakthrough, 43
  deployment plan, 122
  getting firms ready for, 33–34, 80
  IT architecture, 44–47
  knowledge transfer, 34–42, 105, 209
  in large enterprises, 114
  as linking tool, 176
  role of, 150
  supply change progression, 42–43
  technical architecture, 43–44
  tools, 85, 108, 113, 120
  ultimate objective of, 120
  value of, 52–53
Business Process Management Languages
  (BPML) 50, 53, 145, 178
Business process management systems
  (BPMS)
  for customized processes, 108
  getting firm ready for, 80
  information sharing factors, 224–225
  as knowledge link, 48–51, 209
  as linking system, 106–107, 176
  ultimate objective of, 120
  within new architecture, 51–53
Business units
  information transfer across, 180
  leadership, 107
  progressive, 81

C

CAD/CAM, 117
Calibration, 210–213
Call center inquiries, 107

Capacity buffers, 94
Capacity planning, 19
Capital, 63, 72
Capital expenditure, 94
Capital investment, 94, 214
Carrying costs, 63
Cash flow return on investment, 58
CellStar, 122–128, 169–171
CEO, 210, 214
Champy, James, 190
Change
  domains, 85–86, 94–95
  management, 98
  process, 90
  product design process, 17
  team, 95
Changeovers, 94
Channels to market, 89
Churn, 182
Cingular, 123
Cisco Systems, 1, 7, 9
Clarks Shoes, 86–88, 91
Clayton, Raymond, 58
Colgate-Palmolive, 7, 58
Collaboration
  agreed-upon principles, 132, 133, 134
  ally selection, 107–109
  best practices and, 89
  cultural inhibitions and, 128
  defined, 84–85
  electronic, 104
  enterprise transformation, 132, 137
  examples, 103–105, 132
  five-step process, 131–137
  governance model, 132, 133, 135
  in intellectual effort, 81
  inter-enterprise, 150
  internal processes validation, 132, 133,
    136, 137
  levels, 102–103, 131
  process consolidation, 132, 137
  questions and issues, 108
  resistance to, 99
  for reverse logistics, 170
Collaborative commerce, 174
Collaborative network(s), 36–37, 40, 108,
  168

Collaborative planning, forecasting, and
  replenishment (CPFR), 50, 114
Collective assets, 106
Co-managed value chain, 119–120
Commercial aircraft, 96–97
Communication system(s)
  advanced-state supply chain, 118
  disparate, 176
  nonexistent, 99
  in third process step, 59
  See also Information sharing; Internet;
    Knowledge sharing
Competitive advantage
  creating, 36
  maturity model and, 224
  supply chain as, 125
  supply change management, 9
  sustainable, 183
Competitive strategy, 109
Complex business networks, 34–42
Complex business systems, 167–168
Complexity, 109
Computer-aided design (CAD), 117
Computer-aided manufacturing (CAM), 117
Computer Sciences Corporation (CSC),
    128, 174
  automotive industry survey, 114
  e4SM architecture, 150, 151
  supply chain assessment, 221, 223
  survey results, 7–11, 39, 77, 81, 99,
    166, 225
  as systems integrator, 170
Configure product process, 17
Consumer goods providers, 7
Consumer groups, 173
Continuous replenishment programs, 63
Co-planning, 139
Core competencies, 158
Corporate excellence, 174–175
Corporate executive committee, 133
Corporate performance management, 207
Cost containment, 159
Cost of goods sold, 63
Cost reduction
  collaboration and, 105
  level 2 to level 3, 114
  supply chain and, 59–64

CPFR. *See* Collaborative planning, forecasting, and replenishment
CRM. *See* Customer relationship management
Cross-organizational communication, 36
Cross-organizational network, 12–17
Cross-selling, 182
CSC. *See* Computer Sciences Corporation
Cultural barrier, 95–96
  between levels 1 and 2, 5–6
  between levels 2 and 3, 173
  external, 106
  in internal communication system, 182
  overcoming, 81, 95–96, 118
Cultural imperative, 177
Current state mapping, 110–111, 213, 214–221
Customer(s)
  experience, 184–186
  identification system, 182
  information integration, 184, 185
  intelligence, 167, 180–188, 186
  segmentation, 181–183, 186
  specific, 173
  strategic, 184
  values, 165–166, 167
Customer-focused processes, 174
Customer intelligence progression, 173
Customer needs
  call centers, 175
  fulfilling, 92
  individual, 108
Customer relationship management (CRM)
  ASCM and, 181
  collaborative, 175
  contemporary view, 176–179
  operational, 175
  progress, 174
  supply chain progress and, 168
Customer satisfaction
  best possible level, 62
  focus on, 68, 167, 181
  highest possible, 56
  intelligent value chain and, 184
  in level 3, 38
  setting new parameters, 42
Customer service

  improvements, 100
  levels, improved, 115
  mission, 92
  ratings, 74
  targets, 92
Customized processes, 108
Cycle time, 98, 105, 160

D

Data
  domains of change, 85
  errors, 98
  key partner, 150
  management, 167
  sharing, results of, 105–106
  structures, 113
  transfer, 108
DDSN. *See* Demand-driven supply network
De-bottlenecked processes, 94
Deliver process, 16, 97–98
Delivery shortages, 97
Delivery time, 103
Dell, 1, 7, 9, 152–153, 166
Delphi Corporation, 214, 218, 221
Demand
  chain management, 176
  common view, 111
  forecasts, 91, 150
  patterns, 98
Demand-driven supply network (DDSN), 34, 158–159
Demand-pull systems, 24
Deming, W. Edwards, 190
Departmental boundaries, 113
"Descriptive modeling," 182
Desert boot supply chain, 87
Design
  capabilities, 105
  files, 116–117
  process and tooling processes, 17
  product process, 16–17
  studios, 98
  teams, 133
Diageo plc, 120, 122
"Digital cockpit," 206

Direct and indirect spending, 11–12
Distribution, 25
Distribution resource planning system, 93
Document transfer, 107
Domains of change, 85–86, 94–95
Dominance, 119
Downstream activities, 91
Downstream plant, 94
Driving metrics, 57–59
Dynamic simulation, 191
Dynamism, 109

E

EAI. *See* Enterprise application integration
Earnings, 11
Earnings per share, 63, 215
E-business, 141, 145, 153
ebXML, 145
Economic model, 161
Economic value added (EVA), 72, 161, 163, 165
Electronic invoice, 118
Electronic linkages, 120, 155
Enabling systems, 159
Enabling technology, 42–43, 85
End-to-end processing, 18, 37, 42
"End-to-end work," 191
Engineering processes, 16–17, 25–31
Enterprise
    architecture, 51
    integration, 174
    model, 16–17
    -wide resource planning systems, 107
Enterprise application integration (EAI), 50
Enterprise resource planning (ERP), 120, 123–124, 138, 159
E-procurement, 12, 96
Equity, 72
ERP. *See* Enterprise resource planning
E-sourcing tools, 24
EVA. *See* Economic value added
Extended enterprise system, 152
    distribution layer, 152
    hidden values within, 105
    at level 5, 160–165, 169

multiplicity of, 108
    process model of, 13
Extended value chain, 25
External collaboration effort, 107–109
External environment, 12
External view, 173
Extranet, 108, 184. *See also* Internet; Intranet(s)

F

Famurewa, Yomi, 133
FedEx, 123
Fill rates, 68
Financial performance, 72–73
Firewall, 53, 104
"Five big ideas," 190
Fixed asset utilization, 221, 222
Fixed capital, 94
Flextronics, 123
Flow control, 114
Fontanella, John, 35
Ford, 213
Ford, Henry, 190
Forecasting, 62
Forrester survey results, 112
Freight, 3, 11, 105
Fuego, Inc., 122
Fuji, 104
Full network connectivity, 155, 171
Full plant conditions, 93
Functional organization, 97
Functional silos, 89
Future state mapping, 110–111, 213, 214–221

G

Gartner Group Research, 48, 114, 169
Gatekeeper program, 124–125
General Electric, 7, 9
General Motors, 115–117, 213
Gillette Company, 56
Gladiator appliance line, 105
Glass pipeline, 152
Global automotive industry, 213–214
Global network, 102, 167

Goldratt, Eli, 190
Governance model(s), 132, 140, 150, 163, 165
GPS technology, 25
Guinness beer, 120

H

Hammer, Michael, 190
Hard-coded flow control, 114
Harley-Davidson, 117–118
Heuristic algorithm, 199
Hewlett-Packard, 7, 9
High-technology companies, 7
High turnover stock, 170
Hoffman, Debra, 166
Horizontal businesses, 65–66
Horizontal organization, 86, 89–90
Human resources, 159
Hybrid decentralized/centralized model, 96

I

i2, 152
IBM Corporation, 7, 9, 56, 57
Improvement(s)
  in complicated environment, 167–168
  higher level, 156–160
  internal, 82–83
  process, 92, 99–100
  short-term, 81
  teams, 113
  three A's approach, 82–83
Inbound freight, 3, 11
Inconsistent customer experiences, 180
Indifferent thinking, 118
Industry networks, 174
Industry standards, 52
Information, transparent flow, 166
Information sharing
  barriers, 33
  internal resistance, 39, 155
  See also Knowledge sharing
Information technology (IT)
  advanced environment, 160
  in capability matrix, 159

department, 77–78
improvements, 116
infrastructure, 42
investments, 51, 78
network collaboration, 102
new applications, 85
project cancellation, 114
rework, 34
  See also Extranet; Internet; Intranet(s)
Information transfer, 180
Ingram Micro, 123
Initiatives, 94
Innotrac, 123
Insight-driven interactions, 184
Instant messaging, 104
Intalio, 113
Intel, 1, 7, 166
Intelligence sharing, 168
Intelligent network environment, 166
Intelligent value chain, 184, 185
Intelligent value network
  building network business, 179
  for complex business networks, 34
  customer focus, 209
  greater value through, 101
  key ingredients of, 167
  process maturity roadmap and, 13
  solid link into, 166
Inter-enterprise collaboration, 150
Internal cooperation, 84
Internal fragmentation, 181
Internal handling, 11
Internal processes, 80
Internal "silo" processes, 136
Internet
  BPM and, 53
  community, 24
  customer satisfaction and, 181
  as major communication tool, 175
  in product design, 101
  in value management, 161
  See also Communication system(s);
    Information sharing; Intranet(s);
    Knowledge sharing
Intra-enterprise initiatives, 83
Intranet(s), 132, 182. *See also* Extranet

Inventory
  across plants, 93
  across whole chain, 91, 94
  network visibility and control, 62
  nonearning assets, 170
  reduction, 63, 99, 105, 115, 218, 220
  redundant, 74
  RONA and, 93–94
  in supply chain map, 95–96
  turns, 105
  Web-based procurement, 152
  working capital and, 63
Investment(s), 73
  capital, 94
  categories of, 165
  decisions, shared, 155–156
  funding opportunities, 72
  IT, 78, 115
  local, 165
  maximizing value, 73
  payback, 94
  risk, 164
IT. *See* Information technology

J

Johnson & Johnson Medical, Inc., 57
Johnson Controls, 214, 218, 221
Just-in-time purchasing, 63

K

Kanban-type systems, 24
Kawasaki, 104
Key partner data, 150
Key performance indicators (KPIs), 84,
  205–206
Knowledge management, 104
Knowledge sharing
  electronically, 107–108
  IT in, 78
  obstacles, 121, 155
  results of, 105–106
  trusting atmosphere, 106
  upstream, 160
  See also Information sharing

Knowledge transfer, complex networks,
  34–42
Kontzer, Tony, 104
KPIs. *See* Key performance indicators
KYB Manufacturing, 57
Kyocera Wireless, 158

L

Labor costs, 74, 99
Lanner's Witness Simulation package, 200,
  201
Leadership
  advocates and cynics, 82
  forward-thinking, 118
  strong, 102
  visionary, 107
Leading-edge practices, 118
Lead times
  gains, 100, 105
  manufacturing, 98
  in supply chain map, 95–96
Lean manufacturing techniques, 62
Lean production, 190
Legacy systems, 33, 111, 138
Linking processes barriers, 111
Local initiatives, 94
Local performance measurement, 97
Location of work, 85
Lockheed Martin, 102
Logistics, 159. *See also* Reverse logistics
  costs, 11

M

Maintain (manufacturing) process, 17
Make process, 14, 16, 97–98
Manage product portfolio process, 16
Manage Product processes, 98–99
Manual processing, 85, 110, 114
Manufacturing
  companies, 132
  excellence, 159
  lead times, 98
  process, 62
Manugistics Group, Inc., 120, 122

Market
 advantage, 166, 183
 dominance, 119
 leadership, 167, 174
 shifts, 165
Marketing, 180
Mass-market techniques, 180
Material costs, 99
Material requirements planning (MRP), 19,
  97–98
Materials Velocity Centers, 118
Mattel, Inc., 101
Maturity matrices
 design engineering, 26–27
 production engineering, 28
 technology, 30–31
Maturity model
 competitive advantage and, 224
 concepts behind, 2–7
 customer intelligence, 182, 183, 187,
  188
 levels 1 to 5, 2–7, 64
 matching enabling technology, 37–38
 matrices, 18, 20–21
 moving from level 1 to 2, 100
 supply chain, 173–174
Maximum business advantages, 80
McDonnell Douglas, 96
Menlo, 123
Metrics
 financial and nonfinancial, 86
 for judging processes, 68
 for process improvements, 57–59,
  99–100
 value chain, 73
MG Rover Group Ltd., 57
Michlowitz, Eric, 153
Middleware, 50
Minimum batch sizes, 98
Mitsubishi, 104
Modeling, 186, 193
Motorola, 122
Moving production line, 190
MRP. *See* Material requirements planning
Multidivisional corporation, 65
Multiple supply partners, 37
Multitier logistics systems, 25

N

N:1 Advantage©, 130
Nestlé, 102
Net operating profit against capital
  invested (NOPACI), 214
Network
 allies, 173
 collaboration, 101
 connectivity, 121
 extranet, 132
 global, 102
 improvements, 173
 sourcing, 158
 visibility and control, 62
Networked enterprise, 119
New revenues, 64, 168
New technologies, domain of change, 85
Nike, 1, 7, 166
Nonearning assets, 170
Nonfinancial key performance indicators, 68
Nonfinancial measures, 86, 100
NOPACI. *See* Net operating profit against
  capital invested
North American Manufacturing Study, 56
Nucleus firm, 176, 183

O

"Omnigistics," 125, 169–170
On-time delivery, 68
Open standards, 52
Operating margins, 63
Operationalization, 186
Operational management, 190
Optimization
 algorithms, 198–199
 examples, 200–204
 possibilities, 198–200
 types, 200, 205
Optimized conditions, 121
Optimized internal processing, 89
Optimized operating conditions, 85
Order entry and tracking, 105
Order streams, 97
Order-to-cash process, 145
Organizational alignment, 174

Organizational changes, domain of change, 85
Outbound freight, 3, 11
Outsourcing, 105, 110
Overstocking, 103

P

Package processes, 48
Pan-enterprise processes, 72–73
Pan-supply-chain thinking, 95, 96
Partnering, 156
Partners. *See* Business partners; Supply
    chain partners; Supply partners
Parts centers, 118
Payback, 94, 98
PDM. *See* Product data management
Pegasystems, Inc., 122
Pepsico, Inc., 56
Performance
    data, 150
    goals, 95
    measurement, 95, 97
Planning
    across supply chain, 95–96
    communication and, 62
    improved, 63
Plan process, 14, 90–91
Plant capacity, 93
"Predictive modeling," 182
Predictive sourcing, 158
Presence awareness, 104
Process
    customer-focused, lack of, 174
    customization, 108
    design patterns, 52
    difficulties, overcoming, 167
    driving improvements, 114
    explicitly defined, 114
    financial link to, 64–66
    flexibility, 119
    flow, 108
    focus, 190
    improvements, 83, 99–100, 225
    inter-company, 113
    management, 150
    mapping, 100

optimization, 37
optimum, 159
orientation, 181
prioritization, 113
protocols, 145
thinking, 99
time, 117
Process changes
    as domain of change, 85
    value adding, 80–82
Process maturity
    levels, 77
    value of increasing, 73–76
    See also Maturity matrices
Process simulation
    advantages of, 194, 195–196
    defined, 189–190
    discrete event simulation, 191–192, 194,
        196–197
    dynamic simulation, 191
    experimentation advantages, 197
    introduction, 191–193
    key performance indicators and, 205–206
    optimization, 198–199
    for potential business processes,
        193–195
    process focus, 190–191
    reasons for, 193–195
Procter & Gamble, 1, 7, 9, 59, 102, 166
Procurement, 96, 223
Product data management (PDM), 139
Product development process time, 117
Product family
    "dashboard," 86, 88
    owner, 89
    RONA and, 94
    at Syngenta, 91
    unique asset sets of, 86
Profiling tool, 182
Profitability, 170
Profit and loss (P&L) statements, 66–69
    examples, 56–57
    isolated benefits, 84
    supply chain strategy and, 10
Protocols, 145, 150
Pull model, 214
Push model, 214

Q

Quaker Oats, 56
Queues, 194, 196

R

RACAL, 98
Randomness, 192
Random number sampling, 197
Real-time interaction, 104
"Replace and scrap" policy, 18
Reporting lines, 95
Return handling, 169–171
Return on inventory, 74
Return on investment (ROI), 64, 169
Return on net assets (RONA)
  inventory and, 93–94
  maximizing, 89, 91
  model for value chain, 70–71
  in supply chain, 64–65, 91, 92, 95
  at Zeneca Agrochemicals, 71
Return on technology, 115
Return process, 16, 97–98
Revenue
  growth, 215, 217
  improvement, 114
  opportunities, 105
Reverse logistics, 128, 169–171. See also
  Logistics
Rockport, 122
Rockwell, 96
ROI. See Return on investment
Roles, 89
RONA. See Return on net assets
Root cause analysis, 98
Root cause of problems, 99, 221
RosettaNet, 145
Rudzki, Robert, 156
Rules-based processes, 48

S

Safety stocks, 74, 105
Sales and operations planning (S&OP), 4,
  19, 63
SAP, 111

SAP R/3, 95
Savings, 12
Schedule synchronization, 139
Scheduling systems, 24
Scoring technique, 211–213
SCOR® model
  best use of, 122
  enhancements, 210
  five key processes, 13–16, 99–100, 115
  linked supply chain processes and, 59,
    62
  planning, 4
  top processes, 18
Seamless electronic linkages, 120
Seasonal product, 92, 93
Security
  concerns, 107, 108
  risk management, 62
Segmentation, 181–183, 186
Selling, general, and administrative
  (SG&A) costs, 124, 128, 218, 219
Senior management
  alignment, 210
  in automotive industry, 214
  awareness session, 210
  internal improvement and, 82
  partnerships and, 156
Shared protocol, 145
Shared use of equipment and facilities, 11
Shareholder value, 63, 72, 163
Shop floor control, 24
Short cycle manufacture, 94
Short-term improvement, 81
Silo-based planning, 18
Silo mentality, 3
"Silo" processes, 136
Silos, 89
Simulation. See Process simulation
Sinur, Jim, 48
Six Sigma quality, 62, 159
Software
  applications, 114
  browser-based interfaces, 152
  componentized, 42
  development cost, 170
  disparate, 33
  layered model, 108

provider, 108
   team collaboration, 104
Solectron, 123
Source process, 14, 17, 96–97
Sourcing, 105, 110, 158, 223
Spend categorization, 96
Spend reporting, 97
Sprint PCS, 123
SSDW. *See* Strategic sourcing data
   warehouse
Standardized product, 190
Statistical control, 190
Stock buffers, 94
Stock dispositions, 150
Stockless distribution centers, 160
Stock-outs, 97
Storage, 62
Stovepipe mentality, 3, 43, 99, 181
Strachen & Henshaw, 98
Stratascope, 128, 214, 215
Strategic imperatives, 109
Strategic sourcing, 223
Strategic sourcing data warehouse
   (SSDW), 139
Strategic value, 183
"Supernet," 108
Supplier
   management system, 138
   performance, 97
   portal, 117–118
   quality signoff, 97
Supply bases, 3
Supply chain(s)
   barriers to improvement, 111–112
   as collaborative networks, 36–37, 168
   cost reduction and, 59–64
   distinctions in, 64
   end-to-end improvement effort, 166–167
   evolution, 3
   extended, 150
   horizontal, 66
   improvement and, 82, 215
   key business drivers, 209
   levels, 38,100
   linkage, 175
   management, 9
   mapping, 71, 85–86, 95–96, 161, 162

partnerships, 11, 156
   process analysis, 109–111
   progression, 42–43
   roadblocks, 79–80, 95–96
   suggestions, 218
   ultimate benefits, 169
   See also Inventory; Maturity model
Supply-Chain Council, 4, 13
Supply Chain Management Review, 7, 58,
   115
   survey results, 7–11, 39, 77, 79, 81, 99,
      166, 225
Supply-Chain Operations Reference model.
   See SCOR® model
Supply partners, 37
Supply Power, 117
Supporting architecture, 42–43
Swan, Tim, 104
Syngenta, 68, 91
Szygenda, Ralph, 115–116

T

Target, 122
   customer service, 92
Targeted market segments, 180
Targeting tool, 182
Taylor, Frederick, 190
Teal Group, 104
Technology maturity matrix, 22–23
Telecommunications products, 158
Tenneco Automotive, 214, 215, 216
Tesco, 1, 102
Thales, 98
Theory of Constraints, 190
Three A's approach, 82–83
Total enterprise optimization, 36, 85, 159,
   165–168
Total network connectivity, 39
Total value chain, 119
Toyota, 1, 7, 9, 166, 190, 213
Trade settlements, 107
Trading relationships, 115
Traditional benchmarking, 18
Traditional thinking, 99
Transaction process, 163, 164
Transportation costs, 105

Trusting atmosphere, 106
Turnover rate, 205–206

U

U.K., 160
Unilever, 57
UPS, 123
Up-selling, 182
Upstream activities, 91
Upstream plant, 94
Upstream processes, 94
USPS, 123
US Transcom, 122
Utilization, 97

V

Value
  added, 80–82, 113, 196
  assessment steps, 126
  awareness of, 160–161
  collaboration and, 102
  creation, 74–76
  gaps, 221
  management, 160–165, 225
  network, 159
  proposition, 112, 155
  selling, 101
  strategic, 183
Value-based transformation plan, 56
Value chain
  collaboration, 84
  constellation, 176
  intelligent, 155
  lead company, 154
  level 4 characteristics, 152
  metric, 73
  organizations, 150
  partners, 164
  processing, 139
Variability, 194
Vendor-managed inventory, 25, 63

Virtual logistics operations, 25
Virtual logistics optimization, 159
Virtual ownership of assets, 11
"Virtual test bench," 189
Visibility
  electronic, 159
  into linked partners operations, 62
  on-line, 105
  real time, 150
  for reverse logistics, 170
Visionary, 107
Visualization of collaboration, 84
Vought Aircraft Industries, 104

W

Wal-Mart, 1, 9, 102, 166
Warehousing, 105
Warranty programs, 169, 170, 171
Web
  application, 120
  inventory procurement, 152
"What if" analysis tool, 193
Whirlpool, 105
Wireless communications industry, 122–128, 169–171
Working capital, 63, 94

X

XML messaging, 123

Y

Yantra, 122–128, 169–171

Z

Zeneca Agrochemicals
  case illustration, 91–94
  change team, 95
  internal supply chain, 68, 69, 70, 71
  value in supply chain, 163